Visions of the Future: **Physics and Electronics**

Leading young scientists, many holding prestigious Royal Society Research Fellowships, describe their research and give their visions of the future. The articles, which have been re-written in a popular and well-illustrated style, are derived from scholarly and authoritative papers published in a special Millennium Issue of the Royal Society's *Philosophical Transactions* (used by Newton; this is the world's longest-running scientific journal). The topics, which were carefully selected by the journal's editor, Professor J. M. T. Thompson FRS, include quantum physics and its relation to relativity theory and human consciousness, electronics for the future, exotic quantum computing and data storage, telecommunications and the Internet. This book conveys the excitement and enthusiasm of the young authors for their work in physics and electronics. Two companion books cover astronomy and earth science, and chemistry and life science. All are definitive reviews for anyone with a general interest in the future directions of science.

MICHAEL THOMPSON is currently editor of the Royal Society's *Philosophical Transactions* (Series A). He graduated from Cambridge with first-class honours in Mechanical Sciences in 1958, and obtained his PhD in 1962 and his ScD in 1977. He was a Fulbright researcher in aeronautics at Stanford University and joined University College London (UCL) in 1964. He has published four books on instabilities, bifurcations, catastrophe theory and chaos and was appointed professor at UCL in 1977. Michael Thompson was elected FRS in 1985 and was awarded the Ewing Medal of the Institution of Civil Engineers. He was a senior SERC fellow and served on the IMA Council. In 1991 he was appointed director of the Centre for Nonlinear Dynamics.

Visions of the Future:

Physics and Electronics

Edited by J. M. T. Thompson

PUBLISHED BY THE PRESS SYNDICATE OF THE UNIVERSITY OF CAMBRIDGE
The Pitt Building, Trumpington Street, Cambridge, United Kingdom

CAMBRIDGE UNIVERSITY PRESS
The Edinburgh Building, Cambridge CB2 2RU, UK
40 West 20th Street, New York, NY 10011-4211, USA
10 Stamford Road, Oakleigh, VIC 3166, Australia
Ruiz de Alarcón 13, 28014 Madrid, Spain
Dock House, The Waterfront, Cape Town 8001, South Africa

http://www.cambridge.org

First published 2001

Printed in the United Kingdom at the University Press, Cambridge

Typeface Trump Mediaeval 9/13 pt. *System* QuarkXPress™ [SE]

A catalogue record for this book is available from the British Library

Library of Congress Cataloguing in Publication data

Visions of the future : physics and electronics / J.M.T. Thompson
p. cm.
ISBN 0 521 80538 4 (pb)
1. Quantum theory. 2. Electronics. 3. Computers. I. Thompson, J. M. T.

QC174.12.V47 2001
530.12–dc21 00-064219

ISBN 0 521 80538 4 paperback

Contents

Preface

Leading young scientists, writing here in a popular and well-illustrated style, describe their research and give their visions of future developments. The book conveys the excitement and enthusiasm of the young authors. It offers definitive reviews for people with a general interest in the future directions of science, ranging from researchers to scientifically minded school children.

All the contributions are popular presentations based on scholarly and authoritative papers that the authors published in three special Millennium Issues of the Royal Society's *Philosophical Transactions*. This has the prestige of being the world's longest-running scientific journal. It was founded in 1665 and has been publishing cutting-edge science for one third of a millennium. It was used by Isaac Newton to launch his scientific career in 1672 with his first paper 'A new theory about light and colours'. Under Newton's presidency, from 1703 to his death in 1727, the reputation of the Royal Society was firmly established among the scholars of Europe and today it is the UK's academy of science. Many of the authors are supported financially by the society under its prestigious research-fellowships scheme.

Series A of the *Philosophical Transactions* is devoted to the whole of physical science and, as its editor, I made a careful selection of material to cover subjects that are growing rapidly and likely to be of long-term interest and significance. Each contribution describes some recent cutting-edge research, as well as putting it in its wider context and looking forward to future developments. The collection gives a unique snap-shot of the state of physical science at the turn of the millennium, while CVs and photographs of the authors give a personal perspective.

The three Millennium Issues of the journal have been distilled into three corresponding books by Cambridge University Press. These cover

Astronomy and Earth Science (covering the creation of the universe according to the big-bang theory, man's exploration of the solar system, the Earth's deep interior, global warming and climate change), *Physics and Electronics* (this volume) and *Chemistry and Life Science* (covering reaction dynamics, new processes and materials, physical techniques in biology and the modelling of the human heart).

Topics in the present book on physics and electronics include quantum physics and its relation to relativity theory and human consciousness; electronics for the future; exotic quantum computing and data storage; and developments in telecommunications and the Internet.

J. M. T. Thompson, FRS
Editor, *Philosophical Transactions* of the Royal Society,
Centre for Nonlinear Dynamics, University College London

1
Night thoughts of a quantum physicist

Adrian Kent

Department of Applied Mathematics and Theoretical Physics, University of Cambridge, Silver Street, Cambridge CB3 9EW, UK

1.1 Introduction

As the twenty-first century begins, theoretical physics is in a situation that, at least in recent history, is most unusual: there is no generally accepted authority. Each research programme has very widely respected leaders, but every programme is controversial. After a period of extraordinary successes, broadly stretching from the 1900s through to the early 1980s, there have been few dramatic new experimental results in the last fifteen years, with the important exception of cosmology. All the most interesting theoretical ideas have run into serious difficulties and it is not completely obvious that any of them is heading in the right direction. So to speak, some impressively large and well-organised expeditionary parties have been formed and are faithfully heading towards imagined destinations; other smaller and less cohesive bands of physicists are heading in quite different directions. However, we really are all in the dark. Possibly none of us will get anywhere much until the next fortuitous break in the clouds.

I will try to sketch briefly how it is that we have reached this state and then suggest some new directions in which progress might eventually be possible. However, my first duty is to stress that what follow are simply my personal views. These lie somewhere between the heretical and the mainstream at the moment. Some of the best physicists of the twentieth century, would, I think, have been at least in partial sympathy.[1] However, most leading present-day physicists would emphasise different problems; some would query whether physicists can sensibly say anything at all on the topics I will discuss.

I think we can, of course. It seems to me that the problems are as sharply defined as those we have overcome in the past: it just happens that we have not properly tackled them yet. They would be quite untouched – would remain deep unsolved problems – even if what is usually meant by a 'theory of everything' were discovered. Solving them may need further radical changes in our world view, but I suspect that in the end we will find there is no way around them.

1.2 Physics in 1999

The great discoveries of twentieth-century physics have sunk so deeply into the general consciousness that it now takes an effort of will to stand back and try to see them afresh. We should nonetheless try, just as we should try to look at the night sky and at life on earth with childlike eyes from time to time. In appreciating just how completely and how amazingly our understanding of the world has been transformed, we recapture a sense of awe and wonder in the universe and its beauty.[2]

So recall that, in 1900, the existence of atoms was a controversial hypothesis. Matter and light were, as far as we knew, qualitatively different. The known laws of nature were deterministic and relied on absolute notions of space and time that seemed not only natural and common sense but also so firmly embedded in our understanding of nature as to be beyond serious question. The propagation of life and the functioning of the mind remained so mysterious that it was easy to imagine that their understanding might require quite new physical principles. Nothing much resembling modern cosmology existed.

Einstein, of course, taught us to see space and time as different facets of a single geometry. Then, still more astonishingly and beautifully, he

[1] In any case, I am greatly indebted to Schrödinger and Bell's lucid scepticism and to Feynman's compelling explanations of the scientific need to keep alternative ideas in mind if they are even partially successful, as expressed in, for example, Schrödinger (1954), Bell (1987) and Feynman (1965).

[2] We owe this, of course, not to nature – which gives a very good impression of not caring either way – but to ourselves. Though we forget it too easily, that sense is precious to us.

taught us that the geometry of space–time is nonlinear, that matter is guided by the geometry and at the same time shapes it, so that gravity is understood as the mutual action of matter on matter through the curvature of space–time.

The first experiments confirming an important prediction of general relativity – that light is indeed deflected by the solar gravitational field – took place in 1917: still within living memory. Subsequent experimental tests have confirmed general relativity with increasingly impressive accuracy. It is consistent with our understanding of cosmology, insofar as it can be – that is, insofar as quantum effects are negligible. At the moment it has no remotely serious competitor: we have no other picture of the macroscopic world that makes sense and fits the data.

Had theorists been more timid, particle physics experiments and astronomical observations would almost certainly eventually have given us enough clues to make the development of special and general relativity inevitable. As it happens, though, Einstein was only partially guided by experiment. The development of the theories of relativity relied on his extraordinary genius for seeing through to new conceptual frameworks underlying known physics. To Einstein and many of his contemporaries, the gain in elegance and simplicity was so great that it seemed the new theories almost had to be correct.

While the development of quantum theory too relied on brilliant intuitions and syntheses, it was much more driven by experiment. Data – the blackbody radiation spectrum, the photo-electric effect, crystalline diffraction, atomic spectra – more or less forced the new theory on us, first in *ad hoc* forms and then, by 1926, synthesised. It seems unlikely that anyone would ever have found their way through to quantum theory unaided by the data. Certainly, no one has ever found a convincing conceptual framework that explains to us why something like quantum theory should be true. It just is. Neither has anyone, even after the event, come up with a truly satisfactory explanation of what precisely quantum theory tells us about nature. We know that all our pre-1900 intuitions, based as they are on the physics of the world we see around us every day, are quite inadequate. We know that microscopic systems behave in a qualitatively different way, that there is apparently an intrinsic randomness in the way they interact with the devices we use to probe them. Much more impressively, for any given experiment we carry out on microscopic systems, we know how to list the possible outcomes and calculate the probabilities of each, at least to a very good approximation. What we do not fully understand is why those calculations work: we have, for example, no firmly established picture of what (if anything) is going on when we are not looking.

Quantum theory as it was originally formulated was inconsistent with special relativity. Partly for this reason, it did not properly describe the interactions between light and matter either. Solving these problems took several further steps and in time led to a relatively systematic – though still today incomplete – understanding of how to build relativistic quantum theories of fields and, eventually, to the conclusion that the electromagnetic force and the two nuclear forces could be combined into a single field theory. As yet, though, we do not know how to do that very elegantly and almost everyone suspects that a grander and more elegant unified theory of those three forces awaits us. Neither can we truly say that we fully understand quantum field theory, or even that the theories we use are entirely internally consistent. They resemble recipes for calculation, together with only partial, though tantalisingly suggestive, explanations of why they work. Most theorists believe that a deeper explanation requires a better theory, which has perhaps yet to be discovered.

Superstring theory, which many physicists hope might provide a complete theory of gravity as well as the other forces – a 'theory of everything' – is currently the most popular candidate. Though no one doubts its mathematical beauty, it is generally agreed that so far superstring theory has two rather serious problems. Conceptually, we do not know how to make sense of superstrings as a theory of matter plus space–time. Neither can we extract any very interesting correct predictions from the theory – for example, the properties of the known forces, the masses of the known particles, or the apparent four-dimensionality of space–time – in any convincing way.

Opinions differ sharply on whether those problems are likely to be resolved and hence on whether superstring theory is likelier to be a theory of everything or of nothing: time will tell. Almost everyone agrees, though, that reconciling gravity and quantum theory is one of the deepest problems facing modern physics. Quantum theory and general relativity, each brilliantly successful in its own domain, rest on very different principles and give highly divergent pictures of nature. According to general relativity, the world is deterministic, the fundamental equations of nature are nonlinear and the correct picture of nature is, at bottom, geometrical. According to quantum theory, there is an intrinsic randomness in nature, its fundamental equations are linear and the correct language in which to describe nature seems to be closer to abstract algebra than to geometry. Something has to give somewhere, but at the moment we do not know for sure where to begin in trying to combine these pictures: we do not know how to alter either in the direction of the other without breaking it totally.

However, I would like here to try to look a bit beyond the current conventional wisdom. There is always a danger that attention clusters around some admittedly deep problems while neglecting others, simply through convention, habit, or sheer comfort in numbers. Like any other subject, theoretical physics is quite capable of forming intellectual taboos: topics that almost all sensible people avoid. They often have good reason, of course, but I suspect that the most strongly held taboos sometimes resemble a sort of unconscious tribute. Mental blocks can form because a question carries the potential for revolution, in that addressing it thoughtfully would raise the possibility that our present understanding could, in important ways, be quite inadequate: in other words, they can be unconscious defences against too great a sense of insecurity. Just possibly, our best hope of saying something about future revolutions in physics might lie in looking into interesting questions that current theory evades. I will look at two here: the problem of measurement in quantum theory and the mind–body problem.

1.3 Quantum Theory and the Measurement Problem

As we have already seen, quantum theory was not originally inspired by some parsimonious set of principles applied to sparse data. Physicists were led to it, often without seeing a clear way ahead, in stages and by a variety of accumulating data. The founders of quantum theory were thus immediately faced with the problem of explaining precisely what the theory actually tells us about nature. On this they were never able to agree. However, an effective-enough consensus, led by Bohr, was forged. Precisely what Bohr actually believed (and why) remains obscure to many commentators, but for most practical purposes it has hardly mattered. Physicists found that they could condense Bohr's 'Copenhagen interpretation' into a few working rules that explain what can usefully be calculated. Alongside these, a sort of working metaphysical picture – if that is not a contradiction in terms – also emerged. C. P. Snow captures this conventional wisdom well in his semi-autobiographical novel *The Search* (Snow 1934):

> Suddenly, I heard one of the greatest mathematical physicists say, with complete simplicity: 'Of course, the fundamental laws of physics and chemistry are laid down for ever. The details have got to be filled up: we don't know anything of the nucleus; but the fundamental laws are there. In a sense, physics and chemistry are finished sciences'.
>
> The nucleus and life: those were the harder problems: in everything else, in the whole of chemistry and physics, *we were in sight of the end*. The framework was laid down; they had put the boundaries round the pebbles which we could pick up.
>
> It struck me how impossible it would have been to say this a few years before. Before 1926 no one could have said it, unless he were a megalomaniac or knew no science. And now two years later the most detached scientific figure of our time announced it casually in the course of conversation.
>
> It is rather difficult to put the importance of this revolution into words. [. . .] However, it is something like this. Science starts with facts chosen from the external world. The relation between the choice, the chooser, the external world and the fact produced is a complicated one [. . .] but one gets through in the end [. . .] to an agreement upon 'scientific facts'. You can call them 'pointer-readings' as Eddington does, if you like. They are lines on a photographic plate, marks on a screen, all the 'pointer-readings' which are the end of the skill, precautions, inventions, of the laboratory. They are the end of the manual process, the beginning of the scientific. For from these 'pointer-readings', these scientific facts, the process of scientific reasoning begins: and it comes back to them to prove itself right or wrong. For the scientific process is nothing more nor less than a hiatus between 'pointer-readings': one takes some pointer-readings, makes a mental construction from them in order to predict some more.
>
> The pointer-readings which have been predicted are then measured: and if the prediction turns out to be right, the mental construction is, for the moment, a good one. If it is wrong, another mental construction has to be tried. That is all. And you take your choice where you put the word 'reality': you can find your total reality either in the pointer-readings or in the mental construction or, if you have a taste for compromise, in a mixture of both.

In other words, in this conventional view, quantum theory teaches us something deep and revolutionary about the nature of reality. It teaches us that it is a mistake to try to build a picture of the world that includes every aspect of an experiment – the preparation of the apparatus and the system being experimented on, their behaviours during the experiment and the observation of the results – in one smooth and coherent description. All we need to do science (and all we can apparently manage) is to find a way of extrapolating predictions – which, as it happens, turn out generally to be probabilistic rather than deterministic – about the final results from a description of the initial preparation. To ask what went on in between is, by definition, to ask about something we did not observe: it is to ask in the abstract a question that we have not asked nature in the concrete. According to the Copenhagen view, it is a profound feature of our situation in the world that we cannot separate the abstract and the concrete in this way. If we did not actually carry out the relevant observation, we did not ask the question in the only way that causes nature to supply an answer, so there need not be any meaningful answer at all.

We are in sight of the end. Quantum theory teaches us the necessary limits of science. But are we? Does it? Need quantum theory be understood only as a mere device for extrapolating pointer-readings from pointer-readings? *Can* quantum theory be satisfactorily understood in this way? After all, as we understand it, a pointer is no more than a collection of atoms following quantum laws. If the atoms and the quantum laws are ultimately just mental constructions, is not the pointer too? Is not everything?

Landau and Lifshitz, giving a precise and apparently not intentionally critical description of the orthodox view in a classic 1974 textbook on quantum theory, still seem to hint at some disquiet here:

> Quantum mechanics occupies a very unusual place among physical theories: it contains classical mechanics as a limiting case, yet at the same time requires this limiting case for its own formulation.

This is the difficulty. The classical world – the world of the laboratory – must be external to the theory for us to make sense of it; yet it is also supposed to be contained within the theory. Furthermore, since the same objects play this dual role, we have no clear division between the microscopic quantum and the macroscopic classical. It follows that we cannot legitimately derive from quantum theory the predictions we believe the theory actually makes. If a pointer is only a mental construction, we cannot meaningfully ask what state it is in or where it points; so we cannot make meaningful predictions about its behaviour at the end of an experiment. If it is a real object independent of the quantum realm, then we cannot explain it – or, presumably, the rest of the macroscopic world around us – in terms of quantum theory. Either way, if the Copenhagen interpretation is right, a crucial component in our understanding of the world cannot be theoretically justified.

However, we now know that Bohr, the Copenhagen school and most of the pioneers of quantum theory were unnecessarily dogmatic. We are not forced to adopt the Copenhagen interpretation either by the mathematics of quantum theory or by empirical evidence. Neither is it the only serious possibility available. As we now understand, it is just one of several possible views of quantum theory, each of which has advantages and difficulties. It has not yet been superseded: there is no clear consensus now regarding which view is correct. However, it seems unlikely that it will ever again be generally accepted as the one true orthodoxy.

What are the alternatives? The most interesting, I think, is a simple yet potentially revolutionary idea originally set out by Ghirardi, Rimini and Weber (GRW) in 1986, and later developed further by these authors, Pearle, Gisin and several others. According to their model, quantum mechanics has a piece missing. We can fix all its problems by adding rules to say exactly how and when the quantum dice are rolled. This is done by taking the collapse of the wave function to be an objective, observer-independent phenomenon, with small localisations or 'mini-collapses' constantly taking place. This entails altering the dynamics by adding a correction to the Schrödinger equation. If this is done in the way GRW propose, the predictions for experiments carried out on microscopic systems are almost precisely the same, so that none of the successes of quantum theory in this realm is lost. However, large systems deviate more significantly from the predictions of quantum theory. Those deviations are still quite subtle and very hard to detect or exclude experimentally at present, but they are unambiguously there in the equations. Experimentalists will one day be able to tell us for sure whether or not they are there in nature.

By making this modification, we turn quantum theory into a theory that describes objective events continually taking place in a real external world, irrespective of whether any experiment is taking place and whether anyone is watching. If this picture is right, it solves the problem of measurement: we have a single set of equations that gives a unified description of microscopic and macroscopic physics and we can sensibly talk about the behaviours of unobserved systems, irrespective of whether they are microscopic electrons or macroscopic pointers. The pointer of an apparatus probing a quantum system takes up a definite position – and does so very quickly, not through any *ad hoc* postulate, but in a way that follows directly from the fundamental equations of the theory.

The GRW theory is probably completely wrong in detail. There are certainly serious difficulties in making it compatible with relativity – though recent research gives some grounds for optimism here. Nonetheless, GRW's essential idea has, I think, a fair chance of being right. Before 1986, few people believed that any tinkering with quantum theory was possible: it seemed that any change must so completely alter the structure of the theory as to violate some already-tested prediction. However, we now know that it is possible to make relatively tiny changes that cause no conflict with experiment and that by doing so we can solve the deep conceptual and interpretational problems of quantum theory. We know too that the modified theory makes new experimental predictions in an entirely unexpected physical regime. The crucial tests, if and when we can carry them out, will be made not by probing deeper into the nucleus or by building higher-energy accelerators, but by keeping relatively large systems under careful enough control for quantum effects to be observable. New physics could come directly from the large-scale and the complex; frontiers we thought long ago closed.

1.4 Physics and consciousness

Kieślowski's remarkable film series *Dekalog* begins with the story of a computer scientist and his son who share a joy in calculating and predicting, in using the computer to give some small measure of additional control over their lives. Before going skating, the son obtains weather reports for the last three days from the meteorological bureau and together they run a program to infer the thickness of the ice and deduce that it can easily bear his weight. Tragically, however, they neglect the fire a homeless man keeps burning at the lakeside. Literally, of course, they make a simple mistake: the right calculation would have taken account of the fire, corrected the local temperature and shown the actual thickness of the ice. Metaphorically, the story seems to say that the error is neglecting the spiritual, not only in life, but perhaps even in physical predictions.

I do not myself share Kieślowki's religious worldview and I certainly do not mean to start a religious discussion here. However, there is an underlying scientific question, which can be motivated without referring to pre-scientific systems of belief and is crucial to our understanding of the world and our place in it, which I think is still surprisingly neglected. So, to use more scientifically respectable language, I would like to take a fresh look at the problem of consciousness in physics, where by 'consciousness' I mean the perceptions, sensations, thoughts and emotions that constitute our experience.

There has been a significant revival of interest in consciousness lately, but it still receives relatively little attention from physicists. Most physicists believe that, if consciousness poses any problems at all, they are problems outside their province.[3] After all, the argument runs, biology is pretty much reducible to chemistry, which is reducible to known physical laws. Nothing in our current understanding suggests that there is anything physically distinctive about living beings, or brains. On the contrary, neurophysiology, experimental psychology and evolutionary and molecular biology have all advanced with great success, based firmly on the hypothesis that there is not. Of course, no one can exclude the possibility that our current understanding could turn out to be wrong – but, in the absence of any reason to think so, there seems nothing useful for physicists to say.

I largely agree with this view. It *is* very hard to see how any novel physics associated with consciousness could fit with what we already know. Speculating about such ideas *does* seem fruitless in the absence of data. Nonetheless, I think we can say something. There is a basic point about the connection between consciousness and physics that ought to be made, yet seems never to have been clearly stated and suggests that our present understanding almost cannot be complete.

The argument for this goes in three steps. First, let us assume, as physicists quite commonly do, that any natural phenomenon can be described mathematically. Consciousness is a natural phenomenon and at least some aspects of consciousness – for example, the number of symbols we can simultaneously keep in mind – are quantifiable. On the other hand we have no mathematical theory even of these aspects of consciousness. This would not matter if we could at least sketch a path by which statements about consciousness could be reduced to well-understood phenomena. After all, no one worries that we have no mathematical theory of digestion, because we believe that we understand in principle how to rewrite any physical statement concerning digestion as a statement about the local

[3] Penrose is the best-known exception: space does not permit discussion of his rather different arguments here, but see Penrose (1989, 1994).

concentrations of various chemicals in the digestive tract and how to derive these statements from the known laws of physics. However, we cannot sketch a similar path for consciousness: no one knows how to transcribe a statement of the form 'I see a red giraffe' into a statement about the physical state of the speaker. To make such a transcription, we would need to attach a theory of consciousness to the laws of physics we know: it clearly cannot be derived from those laws alone.

Second, we note that, despite the lack of a theory of consciousness, we cannot completely keep consciousness out of physics. All the data on which our theories are based ultimately derive from conscious impressions or conscious memories of impressions. If our ideas about physics included no hypothesis about consciousness, we would have no way of deriving any conclusion about the data and hence no logical reason for preferring any theory over any other. This difficulty has long been recognised. It is dealt with, as best we can, by invoking what is usually called the principle of psycho-physical parallelism. We demand that we should at least be able to give a plausible sketch of how an accurate representation of the contents of our conscious minds could be included in the description of the material world provided by our physical theories, assuming a detailed understanding of how consciousness is represented.

Since we do not actually know how to represent consciousness, that may seem an empty requirement, but it is not. Psycho-physical parallelism requires, for example, that a theory explain how anything that we may observe can come to be correlated to something happening in our brains and that enough is happening in our brains at any given moment to represent the full richness of our conscious experience. These are hard criteria to make precise, but asking whether they could plausibly be satisfied within a given theory is still a useful constraint.

Now the principle of psycho-physical parallelism, as it is currently applied, commits us to seeing consciousness as an epiphenomenon supervening on the material world. As William James magnificently put it (James 1879):

> Feeling is a mere collateral product of our nervous processes, unable to react upon them any more than a shadow reacts on the steps of the traveller whom it accompanies. Inert, uninfluential, a simple passenger in the voyage of life, it is allowed to remain on board, but not to touch the helm or handle the rigging.

Third, the problem with all of this is that, as James went on to point out, if our consciousness is the result of Darwinian evolution, as it surely must be, it is difficult to understand how it can be an epiphenomenon. To sharpen James' point: if there is a simple mathematical theory of consciousness, or of any quantifiable aspect of consciousness, describing a precise version of the principle of psycho-physical parallelism and so characterising how it is epiphenomenally attached to the material world, then its apparent evolutionary value is fictitious. For all the difference it would make to our actions, we might as well be conscious only of the number of neutrons in our kneecaps or the charm count of our cerebella; we might as well find pleasures painful and vice versa. In fact, of course, our consciousness tends to supply us with a sort of executive summary of information with a direct bearing on our own chances of survival and those of our genes; we tend to find actions pleasurable or painful depending on whether they are beneficial or harmful to those chances. Though we are not always aware of vital information (and are always aware of much else) and though our preferences certainly don't perfectly correlate to our genetic prospects, the general predisposition of consciousness towards survival is far too strong to be simply a matter of chance.

Now, of course, almost no one seriously suggests that the main features of consciousness can be the way they are purely by chance. The natural hypothesis is that, since they seem to be evolutionarily advantageous, they should, like our other evolutionarily advantageous traits, have arisen through a process of natural selection. However, if consciousness really is an epiphenomenon, this explanation cannot work. An executive summary of information that is presented to us, but has no subsequent influence on our behaviour, carries no evolutionary advantage. It may well be advantageous for us that our brains run some sort of higher-level processes that use the sort of data that consciousness presents to us and is used to make high-level decisions about behaviour. However, according to the epiphenomenon hypothesis, we gain nothing by being conscious of these particular processes: if they are going to run, they could equally well be run unconsciously, leaving our attention focused on quite different brain activities or on none at all.

Something, then, is wrong with our current understanding. There are really only two serious possibilities. One is that psycho-physical parallelism cannot be made precise and that consciousness is simply scientifically inexplicable. The other is that consciousness is something that interacts, albeit perhaps very subtly, with the rest of the material world rather than simply passively co-existing alongside that world. If that were the case, then we could think of our consciousnesses and our brains – more precisely, the components of our brains described by physics as it is understood at present – as two coupled systems, each of which influences the other. That is a radically different picture from the one we now have, of course. However, it does have explanatory power. If it were true, it would be easy to understand why it might be evolutionarily advantageous for our consciousness to take a particular form. If, say, being conscious of a particular feature of the environment helps to speed up the brain's analysis of that feature, to focus more of the brain's processing power on it, to execute relevant decisions more quickly, or to cause a more sophisticated and detailed description to enter into our memory, then evolution would certainly cause consciousness to pay attention to the relevant and neglect the irrelevant.

We have to be clear about this, though: to propose this explanation is to propose that the actions of conscious beings are not properly described by the present laws of physics. This does not imply that conscious actions cannot be described by any laws. Far from it: if *that* were the case, we would still have an insoluble mystery; and, once we are committed to accepting an insoluble mystery associated with consciousness, then we have no good reason to prefer a mystery that requires amending the laws of physics over one that leaves the existing laws unchallenged. The scientifically interesting possibility – the possibility with maximal explanatory power – is that our actions and those of other conscious beings are not perfectly described by the laws we know at present, but could be by future laws including a proper theory of consciousness.

This need not be true, of course. Perhaps consciousness *will* forever be a mystery. However, it seems hard to justify confidently any *a priori* division of the unsolved problems in physics into the soluble and the forever insoluble. We ought at least to consider the implications of maximal ambition. We generally assume that everything in nature except consciousness has a complete mathematical description: that is why, for example, we carry on looking for a way of unifying quantum theory and gravity, despite the apparent difficulty of the problem. We should accept that, if this assumption is right, it is at least plausible that consciousness also has such a description. This in turn forces us to accept the corollary – that there is a respectable case for believing that we will eventually find that we need new dynamical laws – even though nothing else we know supports it.

One final comment: nothing in this argument relies on the peculiar properties of quantum theory, or the problems it poses. The argument runs through equally well in Newtonian physics. Maybe the deep problems of quantum theory and consciousness are linked, but it seems to me that we have no reason to think so. It follows that anyone committed to the view I have just outlined must argue that a deep problem in physics has generally been neglected for the last century and a half. So let me try to make that case.

There is no stronger or more venerable scientific taboo than that against enquiry, however tentative, into consciousness. James, in 1879, quoted 'a most intelligent biologist' as saying:

> It is high time for scientific men to protest against the recognition of any such thing as consciousness in scientific investigation.

Scientific men and women certainly have made this protestation, loudly and often, over the last hundred and twenty years – but have those protests ever carried much intellectual force?

The folk wisdom, such as it is, militating against the possibility of a scientific investigation of consciousness seems now to rest on a confusion hanging over from the largely deleterious effect of logical positivism on scientists earlier this century. Hypotheses about consciousness are widely taken to be *ipso facto* unscientific because consciousness is at present unmeasurable and its influences, if any, are at present undetectable. If we delete 'at present', the case could be properly made; as it is, it falls flat. But if logical positivism is to blame, it is only the most recent recruit to the cause. The problem seems to run much deeper in scientific culture. Schrödinger (1954) described the phenomenon of

> [. . .] the wall, separating the `two paths', that of the heart and that of pure reason. We look back along the wall: could we not pull it down, has it always been there? As we scan its windings over hills and vales back in history we behold a land far, far, away at a space of over two thousand years back, where the wall flattens and disappears and the path was not yet split, but was only *one*. Some of us deem it worthwhile to walk back and see what can be learnt from the alluring primaeval unity.

> Dropping the metaphor, it is my opinion that the philosophy of the ancient Greeks attracts us at this moment, because never before or since, anywhere in the world, has anything like their highly advanced and articulated system of knowledge and speculation been established *without* the fateful division which has hampered us for centuries and has become unendurable in our days.

Clearly, the revival of interest in Greek philosophy that Schrödinger saw did not immediately produce the revolution for which he hoped. None the less, our continuing fascination with consciousness is evident on the popular science and philosophy bookshelves. It looks as though breaking down the wall and building a complete worldview are going to be left as tasks for the third millennium. There could hardly be greater or more fascinating challenges.

Neither can there be many more necessary for our long-term well-being. Science has done us far more good than harm, psychologically and materially. However, the great advances we have made in understanding nature have also been used to support a worldview in which only what we can now measure matters, in which the material and the external dominate, in which we objectify and reduce ourselves and each other, in which we are in danger of coming to see our psyches and our cultures, in all their richness, as no more than the evolutionarily honed expression of an agglomeration of crude competitive urges.

To put it more succinctly, there is a danger, as Václav Havel put it in a recent essay (Havel 1996), of man as an observer becoming completely alienated from himself as a being. Havel goes on to suggest that hopeful signs of a more humane and less schizophrenic worldview can be found in what he suggests might be called postmodern science, in the form of the Gaia hypothesis and the anthropic principle.

I disagree: it is hard to pin down precise scientific content in these ideas and, insofar as we can, it seems to me that they are no help. However, I think we have the answer already. The alienation is an artefact, created by the erroneous belief that all that physics currently describes is all there is. However, on everything we value in our humanity, physics is silent. Insofar as our understanding of human consciousness is concerned, though we have learned far more about ourselves, we have learned nothing for sure that negates or delegitimises a humane perspective. In that sense, nothing of crucial importance has changed.

1.5 Postscript

All this said, of course, predicting the future of science is a mug's game. If, as I have argued, physics is very far from over, the one thing of which we should be sure is that greater surprises than anything we can imagine are in store. One prediction that seems likelier than most, though, is that anyone producing a similar volume in the year 3000 will not need to consider only human contributors. Perhaps our future extraterrestrial or electronic colleagues will find some amusement in our attempts. I do hope so.

1.6 Further reading

Bell, J. S. 1987 *Speakable and Unspeakable in Quantum Mechanics: Collected Papers on Quantum Philosophy*. Cambridge: Cambridge University Press.

Feynman, R. 1965 *The Character of Physical Law*. London: British Broadcasting Corporation; and Reading: Addison Wesley.

Havel, V. 1996 in *The Fontana Postmodernism Reader*, ed. W. Truett Anderson. London: Fontana.

James, W. 1879 Are we automata? *Mind* **13** 1–22.

Kent, A. 1998 Quantum Histories. *Physica Scripta* **T76** 78–84.

Kent, A. 2000 Night Thoughts of a Quantum Physicist. *Phil. Trans. Roy. Soc.* A **358** 75–88.

Penrose, R. 1989 *The Emperor's New Mind: Concerning Computers, Minds, and the Laws of Physics*. Oxford: Oxford University Press.

Penrose, R. 1994 *Shadows of the Mind: A Search for the Missing Science of Consciousness*. Oxford: Oxford University Press.

Schrödinger, E. 1954 *Nature and the Greeks*. Cambridge: Cambridge University Press.

Snow, C. P. 1934 *The Search*. London: Victor Gollancz.

2
Metals without electrons: the physics of exotic quantum fluids

Derek K. K. Lee[1] **and Andrew J. Schofield**[2]

[1] *Blackett Laboratory, Imperial College, Prince Consort Road, London SW7 2BW, UK*
[2] *School of Physics and Astronomy, University of Birmingham, Edgbaston, Birmingham B15 2TT, UK*

2.1 Quantum complexity – the new frontier

The close of the nineteenth century and the close of the twentieth exhibit a number of striking parallels. At the turn of the twentieth century, there was a feeling among some theoretical physicists that almost all that could be known was known and all that remained was details to be cleared up. These details appeared to be a number of minor discrepancies between the theory of classical physics and experiment. Classical physics had triumphed with the unification of electricity and magnetism; and J. J. Thomson's discovery of the electron had allowed him to develop a workable version of atomic physics. With the benefit of hindsight we now know that some of those 'minor discrepancies' held the key to a revolution that transformed physics in the 1920s and has dramatically affected the technology of the twentieth century.

That revolution was the discovery of quantum mechanics. It utterly revised our perception of nature at the atomic level. The nineteenth-century view of physics (Newton's mechanics and Maxwell's electromagnetism) proved to be inadequate on these small scales. The pace of development since those days has proved so dramatic that the control of Thomson's electron at a quantum level is now possible. 'Designer electronics' is commonplace in semiconductor technology from televisions and computers to mobile communications – quantum physics is working for us every day.

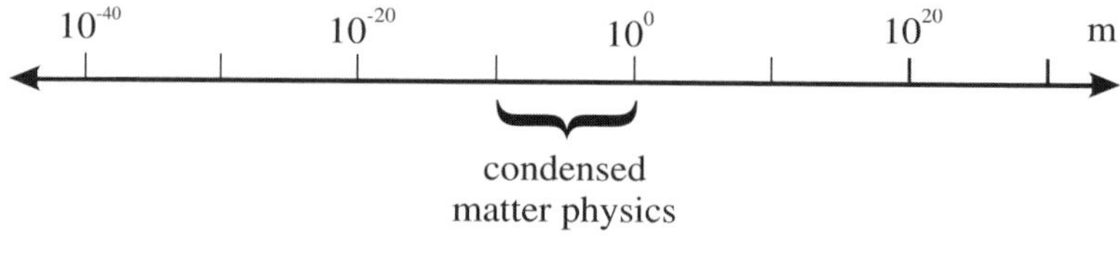

string theory

cosmology

Figure 2.1. It is often tempting to view the challenges of theoretical physics in terms of the extremes of length. Condensed-matter physics concerns itself with distances ranging from about 1 m down to about one ten millionth of a millimetre. In comparison with string theory at the very smallest lengths and cosmology at the size of the universe, condensed matter can seem rather mundane. However, condensed-matter physicists explore a new frontier – extreme complexity (see Figure 2.2).

The successes born of physics in the early years of the twentieth century, quantum mechanics and Einstein's relativity, have left us with the perception that the challenges of theoretical physics lie at the extremes of length scale (see Figure 2.1). At one end, the study of the atom and quantum mechanics has progressed to the challenge of understanding how sub-nuclear particles interact at the smallest imaginable distances. Here, the excitement lies in the realm of string theory and exotic supersymmetries. At the other extreme, on the astronomical scale, the physics of black holes and the beginnings of the universe prove equally fascinating. Mirroring the end of the nineteenth century, some scientists feel that theoretical physics is almost complete. Indeed, the noted cosmologist Stephen Hawking entitled his 1979 inaugural lecture in Cambridge 'The end of theoretical physics?'. He concluded then (as now!) that the end was twenty years away.

However, there is a new dimension emerging in physics, one that explores the physical world from an entirely different perspective. Measured on this alternative axis, we see that theoretical physics is far from over. On the contrary, there are the tell-tale signs that a completely new understanding is required in order to make progress here. Just as at the end of the nineteenth century, there is a growing number of 'discrepancies' between theory and experiments, indicating that there is something important to be discovered. This is a challenge every bit as fundamental to the physics at the extremes of length scale. It is a frontier that has implications for the technology of the twenty-first century, perhaps implications as great as those that have driven the semiconductor revolution.

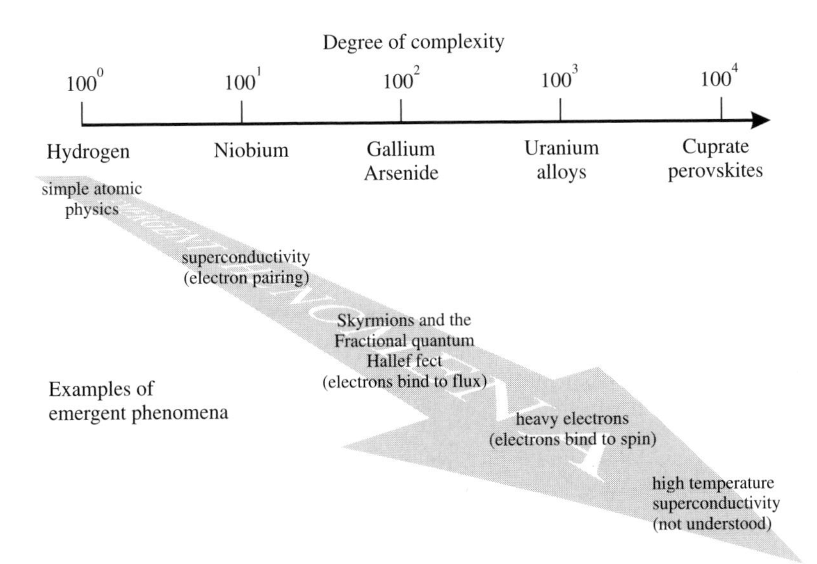

Figure 2.2. Condensed-matter physics can be viewed as the exploration of quantum complexity. Every new element we introduce into a material's structure multiplies by 100 (roughly the number of elements) the number of compounds and hence the complexity. With every new level of complexity we see new properties appearing and their explanations can often require a very different viewpoint from that suggested by the basic underlying ingredients of the electron and its quantum mechanical interactions.

Those rising to this challenge are condensed matter physicists together with materials scientists. The frontier we explore is that of extreme complexity (see Figure 2.2).

Condensed-matter physics is the study of the atomic and electronic behaviours of the everyday matter around us – solids, liquids and gases. The laws of electromagnetism and quantum theory, which govern the behaviours of the electron and atoms, the basic ingredients of this matter, were developed in the 1920s. However, the tremendous advances in materials science have brought these ingredients together in new and complex ways. The new physics emerging from complex systems has shown us that merely knowing the basic constituents and their interactions can often be a poor guide to their collective properties. What we discover in more complex materials is rarely a simple extrapolation of the physics of simpler systems. Instead, we often find entirely new types of behaviour; and we need new concepts to understand them. These 'emergent phenomena' (i.e. newly independent areas of physics) can be seen as we explore the axis of complexity (Figure 2.2). We may loosely parameterise the degree of complexity by the number of possible combinations of materials as we mix together more and more of the hundred or so elements of the periodic table. From the physics of the hydrogen atom we move in complexity to simple metals like niobium. There we see electrons pairing to form a superconductor. Mixing two elements together gives us, for example, gallium arsenide. In devices made from this we see the physics of the fractional quantum-Hall effect, whereby electrons bind to magnetic flux. The tertiary alloys of uranium often appear to be made of electrons with masses thousands of times greater than normal. The current state of the art in materials science lies in the quaternary oxide materials including the 'high-temperature' superconductors. None of these discoveries could have been predicted on the basis of simpler systems. The science behind all of these emergent phenomena shows us that the whole is frequently greater than the sum of the parts.

Understanding how new physical phenomena emerge from complex systems often requires radically new concepts. One dramatic realisation of this forms the subject of this article. Here, we focus on metallic solids in which the interactions between the electrons can lead to strong correlations in their motion. As a consequence, we find exotic quantum phases of matter, for which consideration of *new* types of particles (such as fractionally charged objects) provides a much better basis for their description than does that of the electron. We start by reviewing some of the basic concepts behind our current understanding. We will then describe recent discoveries that appear to demand an entirely new conceptual framework for the theory of metals. These discoveries not only pose a challenge at a theoretical level but also point to opportunities on a technological front.

2.2 The electron fluid: a quantum liquid

An important cornerstone in condensed-matter physics is the theory of metals. Metals, such as copper, conduct electricity. In other words, electrical currents flow from one end of a piece of metal to the other when, for instance, a nine-volt battery is connected across it. The amount of current that flows depends on the material in question. The current would be large in 'good' metals (those with low resistance) but small in 'bad' metals (those with high resistance). To understand the mechanisms that cause resistance, we need to have a microscopic model of a metal. What are these mobile charges? How do they flow? The search for answers to these questions takes us into the realm of quantum mechanics.

The electron: charge and spin

The discovery of the electron marks the beginning of modern physics. By trying to understand its behaviour, we discovered quantum mechanics. The electron is a deceptively simple object. It carries an electrical charge e and is therefore responsible for all things electrical. The unit of charge, e, is fundamental in the sense that the electron cannot be split and that no other particle with a smaller charge has been isolated. However, this does not mean that there are no physical objects with a fraction of an electronic charge. Fractional charge does appear in the context of the fractional quantum-Hall effect (see later).

The electron also possesses a magnetic moment or *spin*. In other words, it behaves in many ways like a tiny bar magnet. For instance, when an electron is placed between the poles of a horseshoe magnet, it will try to align its moment in the same direction as the magnetic field generated by the magnet, that is, towards the south pole. The important difference lies in the quantum nature of the spin – there are only two independent spin states corresponding to opposite directions of spin (frequently referred to as 'up' and 'down'). In a magnetic material, such as iron, as is found in a bar magnet, these spins align parallel with each other, forming a large total magnetic moment. This is a ferromagnet. There are also antiferromagnets, in which neighbouring spins align in opposite directions. One of these materials gives rise to the class of high-temperature superconductors (which will also be discussed later).

Shortly after the discovery of the electron, the first progress in understanding the difference between metals and insulators was made. In an insulator like rubber, the electrons are tied down to specific atoms or molecules and are not free to move around. Metals, on the other hand, should be viewed as a collection of atoms each of which has given up, on average, one or two electrons to be shared among the other atoms. These electrons can conduct electricity because they are now free of their parent atoms and are completely mobile. This 'sea' of free electrons is quite different from any classical gas of particles like the helium gas used to fill balloons.

Helium atoms at room temperature move much like randomly colliding snooker balls governed by Newton's laws of motion. However, we know from atomic physics that the motion of electrons cannot be fully understood through these laws of classical mechanics, for otherwise negatively charged electrons would quickly fall into the positively charged nucleus. In order to understand electrical conduction in metals, we have to take seriously the fact that these electrons form a *quantum* fluid.

In quantum mechanics, electrons live in 'quantum states' (see Figure 2.3). A central principle governing the physics of electrons is the Pauli exclusion principle – two electrons cannot be in the same quantum state. Electrons prefer to arrange themselves so that their total energy is as low as possible. However, if one electron is already occupying a low-energy state, another one cannot adopt the same state and hence is forced to occupy one with higher energy. Some of these electrons end up having very high energies indeed (see Figure 2.3). In a typical metal such as copper, an electron has an average speed of 10^5 m s^{-1}. In fact, this is fast enough for an electron to be shot into space escaping the gravitational pull of the Earth. The reason that this does not happen is that the negatively charged electrons are held back by the strong electrostatic attraction from their parent atoms (which are now positively charged 'ions' after losing an electron to the conduction sea).

We can now build a physical picture of electrical conduction in metals. These fast electrons are accelerated by the electrical field that directs them from one pole of a battery to the other. This is analogous to a ball rolling down a slope, except that the slope is of electrical origin. An electron can bump into impurities and the positively charged ions in the metal. These can deflect the particle away from the direction it is supposed to be following under the influence of the electrical field. It could even be bounced backwards. This is now analogous to a ball falling through a pinball machine. The electrons are hindered in their forward motion by these obstacles so that they do not accelerate indefinitely. They settle down to a steady speed and so we observe a steady current for a given applied voltage. This scattering is the origin of electrical resistance (which is the voltage divided by the current – Ohm's law).

What about electrons getting in each other's way? After all, they repel each other because they have the same charge, so they must scatter each other as well when their paths come close together. It turns out that this mechanism is relatively unimportant. Part of the reason is that, when two

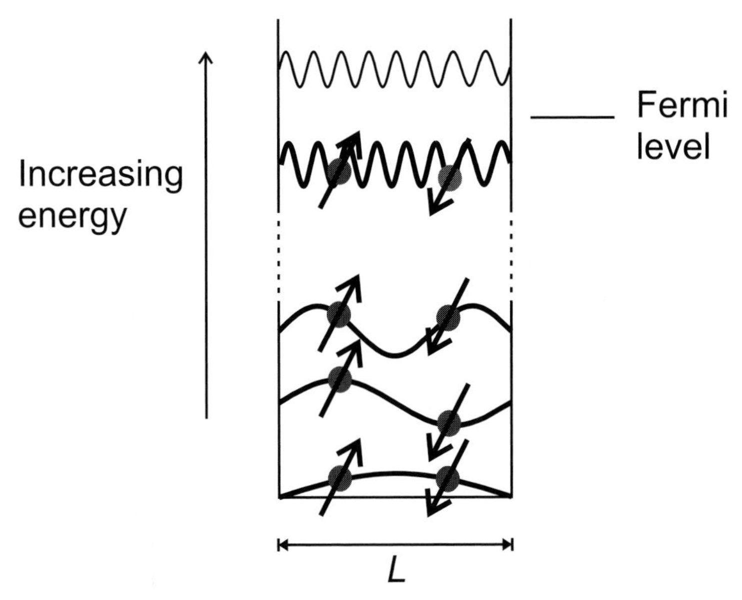

Figure 2.3. Particles in a box. In quantum theory, particles like electrons can also be described as waves trapped by the sides of the material in which they reside. These are like the standing waves that can be formed by jiggling a rope tied to a tree or the vibrations on a violin string. The momentum of the particle, p, is related to the wavelength, λ, by de Broglie's relation $p=h/\lambda$ (where h is Planck's constant). Only specific wavelengths are allowed if we are to fit the standing waves between the end walls of the solid: $\lambda=2L/N$, where N is a whole number and L is the separation of the walls. We say that this *quantises* the momentum: $p=Nh/(2L)$. The system favours the configuration of minimum total energy, the 'ground state'. For one electron, this is the state with lowest momentum. For many electrons, we have to take into account the Pauli exclusion principle, which states that only two electrons (with opposite spins) can be in the same standing-wave state. If we assume that the electrons in a metal do not interact with each other, then it is easy to determine the ground-state configuration. Working upwards in energy, we fill up the available standing-wave states one by one with two electrons each. To accommodate all the electrons (typically 10^{23}), we will reach waves of very short wavelengths, i.e. high momenta and therefore high energies. In a real material, the wave with the highest energy will have a wavelength as short as inter-atomic distances. This top energy level is called the 'Fermi level' and this configuration for the electrons is called the 'Fermi sea'.

electrons scatter off each other, one typically becomes faster and the other slower. As we have already emphasised, other electrons already occupy the states with lower energy (and speed). The Pauli exclusion principle therefore forbids these scattering events!

Nevertheless, the fact that an electron has to swim through a sea of other electrons does have consequences for the properties of the electron. The electrons appear to become heavier in the sense that they are harder to accelerate. We say that the charge carriers appear to be electrons with an 'effective mass' that is larger than the free-electron mass. This can be a significant effect in cases in which electron–electron interactions are strong – the effective mass can be hundreds of times larger than expected. An intuitive picture for this effect is that the electron must push aside other electrons as it moves. We can say that the electron carries with it a positive charge cloud (that is, the space it has created by pushing aside negatively charged electrons.) The Russian physicist Landau coined the term *quasi-particle* for this electron-like object, to distinguish it from the free electron.

This description of electrical resistance essentially assumes that the system behaves like a collection of electron-like objects that do not see each other. This concept is central to our understanding of everyday metals that we find in copper wires and semiconductor chips. The idea is known as Fermi-liquid theory and was formulated by Landau in the 1950s. The success of Fermi-liquid theory is remarkable, given that electrostatics gives strong repulsive forces between electrons. However, these electrons are travelling at high speed. Their paths come close to each other only fleetingly, so their motion is not drastically affected by these brief encounters. We have seen that the high electron velocities arise from the need to accommodate the Pauli exclusion principle. The electron fluid is therefore a system ruled by the laws of quantum mechanics.

The concept of the quasi-particle lies at the heart of Fermi-liquid theory. It is important to note that they are not just theoretical constructs – they are the 'elementary particles' of condensed matter physics since they determine all the physical properties of the system. To borrow a phrase from high-energy physics, Fermi liquid theory has become the 'standard model' of metals. It explains why a wide range of compounds made up of elements from different parts of the periodic table might behave in a similar way. However, it is by no means a 'grand unified theory'. There are many exceptions to the rule that point us towards novel phenomena. Indeed, they challenge us to find a new conceptual framework for the theory of metals. As hinted in our introduction, we are compelled to look for new 'elementary particles' in the quantum fluid, which are radically different from the free electron.

2.3 Beyond Fermi liquids

Some materials behave like an ordinary metal at room temperature but exhibit a completely different behaviour at low temperatures. We say that the system has a transition from a metallic phase to a non-metallic phase. This transition occurs at a very precise temperature, in the same way as that in which water turns into ice at precisely 0 °C. This is a clear case of the breakdown of Fermi-liquid theory.

To give a concrete example, some metals become superconducting at low temperatures: below −266 °C for lead and below −250 °C for niobium germanium (Nb_3Ge). Above the transition temperature, the material is an ordinary metal. Below it, there is no sign of electrical resistance at all. The material also becomes a strong diamagnet, expelling magnetic flux from its interior (the Meissner effect). When magnetic flux is forced into the material with a strong magnetic field, it enters as thin lines, each carrying a *quantized* amount of flux, $h/(2e)$.

What has happened? Clearly, the system has undergone some radical re-organisation. In this case, the electrons (or, more precisely, the electron-like quasi-particles) have paired up. The origin of the electron pairing was explained by Bardeen, Cooper and Schrieffer in 1957 as the result of an attractive interaction mediated by the sound waves in the solid. It is one of the most successful theories in condensed-matter physics.

The new entities (Cooper pairs) belong to a different class of particles from electrons. Like photons, which are particles of light, they are not subject to the Pauli exclusion principle. Instead, they do exactly the opposite – they *prefer* to occupy the same quantum state. They behave as if they are a single collective entity. If you want to deflect them, you have to deflect them *all* at the same time! This cannot be done by single point-like impurities (as is the case of ordinary metals) and so this quantum fluid does not display any electrical resistance. There is in fact an analogous phenomenon in optics: the laser. The laser beam consists of photons acting as a single entity.

In this example, the Fermi liquid breaks down at low temperatures, giving way to a new state of matter. Other examples include ferromagnetism (for iron, below 773 °C), in which the electrons align their magnetic directions spontaneously without the need for an external magnetic field. However, much of the recent interest is in materials in which we see a metallic state with a low (but not zero) resistance, but very different from the one that Landau envisaged. We have a growing number of examples of metals that do not appear to be describable using electrons (or electron-like quasi-particles.) These metals pose a great challenge to our conventional views of metallic, superconducting and magnetic behaviour. Among them, the materials with the greatest technological importance are arguably the cuprate superconductors, the metallic state of which (which exists at temperatures above the superconducting transition temperature) has been the subject of controversy ever since its discovery in 1987.

2.4 The mystery of the cuprate superconductors

The cuprates encompass around thirty distinct crystalline structures and contain upwards of three different elements. It is perhaps not surprising that new phenomena that are not found in simple solids like simple copper emerge in these compounds. Whereas for other metals the onset of superconductivity typically occurs at tens of degrees above absolute zero (−273 °C), the cuprates have transition temperatures around 100 degrees above absolute zero (for example, −183 °C for yttrium barium copper oxide, $YBa_2Cu_3O_7$). The practical significance is that this is above the boiling point of nitrogen (−196 °C) and so this inexpensive liquid can be used as a refrigerant instead of liquid helium. Moreover, the existence of the cuprates raises hopes for the discovery of room-temperature superconductors. These hopes will remain a major driving force behind superconductivity research in physics, chemistry and material science.

In spite of a decade of intensive research, we still do not have a good theoretical understanding of the cuprates. This is because there is an interplay of interactions that influence the physics of the electron system. However, we are coming to the consensus that the answer must lie in a common feature of all the cuprates – the copper and oxygen atoms are arranged in layers. The other atoms act as a sort of scaffolding keeping these layers apart (see Figure 2.4). In each of these layers, the copper atoms are arranged on the corners of a square lattice. The oxygen atoms sit between neighbouring copper sites. In the pristine state of, say, lanthanum cuprate (La_2CuO_4), each copper atom donates one electron as a charge carrier. These electrons prefer to run around within each layer, only cross-

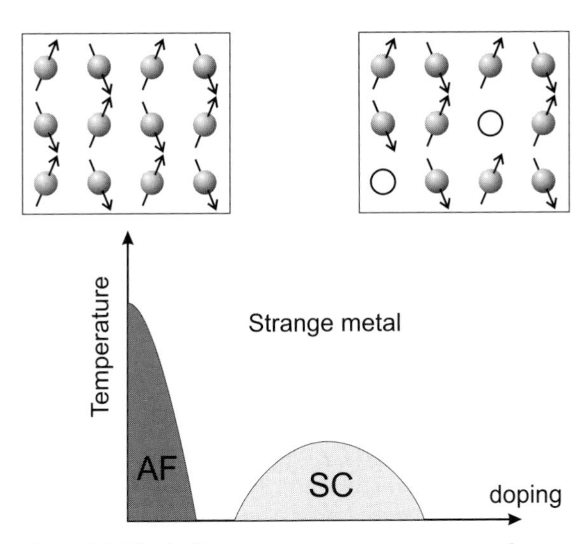

Figure 2.4. The high-temperature cuprate superconductors exhibit a wide range of properties outside those expected for conventional metallic behaviour. A typical material (lanthanum cuprate is illustrated here) has layers of copper and oxide ions. In their pristine state, these layers are usually insulators with an antiferromagnetic arrangement of spins (AF). They are made superconducting by removing electrons from the layers – a process known as doping. This makes the electrons mobile, destroys the magnetism and creates a superconducting material (SC). The metallic state from which superconductivity emerges is so unusual that it is often simply called the 'strange metal phase'.

ing from one layer to the next very occasionally. A first guess for a model describing this system would be to consider electrons hopping from site to site on a square lattice. The two-dimensional nature of this motion is believed to play an important role in the behaviour of this system. This has, in part, prompted a search for unusual materials with low dimensions.

Although a non-interacting electron theory would predict it be a metal, La_2CuO_4 is an insulator. The reason is that the repulsion between electrons is important here, in that it has caused an electronic traffic jam. We have one electron per copper site and an electron cannot hop to a neighbouring site because of the repulsion from the electron that is already there. A system exhibiting this kind of insulating behaviour is called a Mott insulator. Mott insulators are expected to be antiferromagnets – the magnetic moment on copper alternates in direction as one moves from one site to its neighbour. This is observed in all the cuprates.

To make these insulators conduct electricity, we have to relieve the traffic jam. We can do this by changing the chemical composition of the compound so that electrons are transferred from the copper–oxide layers to the atoms residing between the layers. (For La_2CuO_4, we can introduce strontium, which substitutes for the lanthanum atoms.) In this way, electron vacancies (holes) can be created in the layers. The electrons can now move by hopping into the vacancies. Another way of looking at this is to watch the vacancies being shunted around the lattice by the motion of the electrons. The holes have a positive charge relative to the original insulating compound and so we can attribute the conductivity to mobile, positively charged holes.

As we increase the number of electron vacancies, antiferromagnetism soon disappears and the material becomes a metal. This metal becomes superconducting at low temperatures (Figure 2.4). The transition temperature reaches a maximum of about $-170\,°C$ when the lattice contains 15 per cent–25 per cent vacancies. (The record is $-120\,°C$ for a similar compound involving unfortunately toxic mercury.) One might therefore argue that antiferromagnetism and superconductivity compete here. On the other hand, the superconductivity is intimately related to the proximity of the conductor to a magnetic state – the superconductivity disappears at higher concentrations of vacancies when we are far from the original antiferromagnetic insulator.

Leaving aside the superconducting state, the metallic state of the cuprates is itself unusual. One of the earliest observed anomalies is the linear proportionality between the resistance for currents in the copper–oxygen planes and the temperature. This is a robust phenomenon that occurs for nearly all the cuprates when the doping is optimal and it holds over a wide range of temperatures, from the superconducting transition to $700\,°C$. Moreover, the origin of this resistance apparently involves only the electrons. This is surprising because no known theory predicts this behaviour from electronic scattering alone. In fact, Fermi liquid theory expects the resistance to be proportional to the *square* of the temperature, through the scattering of one quasi-particle by another.

We can also measure the rates of scattering of quasi-particles by observing the sideways current generated by the Lorentz force on the electrons due to a magnetic field (the Hall effect). Curiously, the resistance for these currents *does* give a rate of scattering that is proportional to the square of the temperature. How can a single quasi-particle have two radically different rates of scattering? The answer to this question and explanations for many other peculiarities of the cuprates remain highly controversial. Even the co-authors here have different proposals!

As we hinted above, a theoretical model would have to give a quantitative description of how the mobility of the electronic charge and the antiferromagnetic correlations of the electronic magnetic moments are intertwined. Unfortunately, even the simplest theoretical models that include these effects are mathematically intractable beyond one dimension. At present, it is also not known how such a model would give rise to a mechanism for superconductivity.

Nevertheless, many theorists are coming to the view that the experimental anomalies point to a total breakdown of the Fermi liquid in the metallic state of the cuprates. In other words, the fundamentally current-carrying object might be radically different from the electron or any electron-like quasi-particle. The ingredient essential to this system appears to be the presence of some magnetic correlations.

One fruitful direction of research is to broaden our perspective by considering whether a larger range of magnetic materials could exhibit similar non-Fermi-liquid behaviour. In fact, the study of the magnetic metals and insulators pre-dates the discovery of the cuprate superconductors. These compounds exhibit a wide range of behaviour, such as ferromagnetism, antiferromagnetism and superconductivity. For instance, cerium palladium silicon ($CePd_2Si_2$) is an antiferromagnetic metal at atmospheric pressure. The antiferromagnetism can, however, be destroyed at high pressures and the material becomes superconducting on the disappearance of magnetic order (see Figure 2.5). The superconducting critical temperature is very low (0.4 K), but the apparent competition between antiferromagnetism and superconductivity is qualitatively similar to that in the cuprates. This compound might shed some light on how the Fermi-liquid state evolves towards a transition whereby a new quantum state of matter emerges. In particular, it belongs to a class of materials known as 'heavy-fermion compounds' for which a Fermi-liquid interpretation would deduce a quasi-particle that has a mass hundreds or even thousands of times greater than that of the electron. We can regard this as a signal that Fermi-liquid theory has been stretched to the limits of its applicability in considering these materials. The elucidation of the actual nature of the breakdown of the Fermi-liquid state remains a challenge for the future.

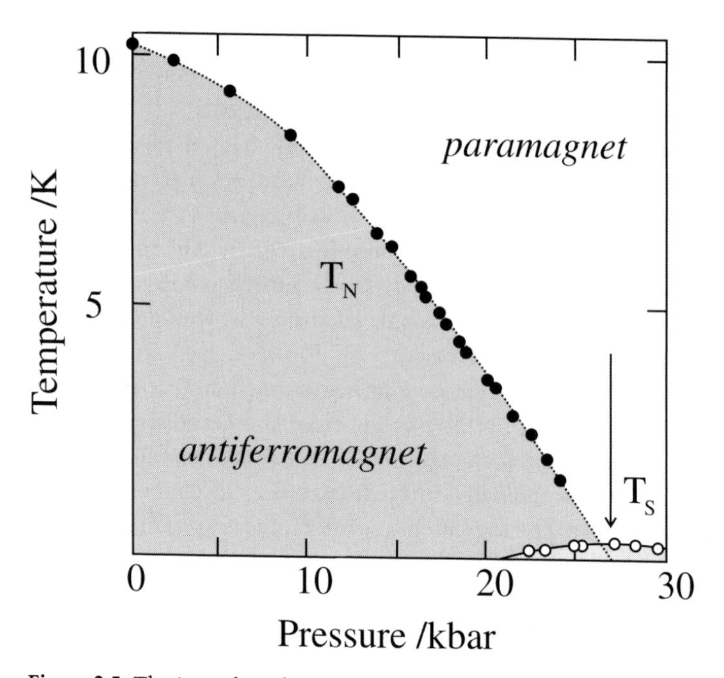

Figure 2.5. The interplay of magnetism, superconductivity and metallic behaviour can be studied by applying pressure as well as by doping as in the case of the cuprates. This has the advantage of minimising the role of disorder. In this material (cerium palladium silicon, $CePd_2Si_2$, studied by Julian and co-workers in Cambridge), the magnetism can be squeezed out of the metal at high pressure (28 000 atm). Near the point where the magnetism is lost, electrons interact strongly with the incipient magnetic order and the whole notion of an electron seems to break down. The theory describing what happens here does not conclusively match the known experimental results. Tantalisingly, at this very point on the phase diagram (indicated by the arrow), superconductivity appears.

2.5 Novel particles?

If we do not have a Fermi liquid, we need to look for a new way of characterising the system. Does this mean that the system is now so complex that it defies description by any simple physical picture? Our hope is that even complex systems have organised ground states and that these quantum states of matter can be characterised by invoking novel quasi-particles. We describe below a few examples in which non-electron-like quasi-particles are known to exist. Notice that, in addition to strong interactions, all these systems involve physics in an effectively reduced number of spatial dimensions.

2.5.1 Spin–charge separation

One conjecture, put forward by Philip Anderson, for understanding the metallic state of the cuprates is that charge and spin (magnetic moment) have somehow become de-coupled in this system. Whereas charge and spin were tied together on the same object (the electron or the electron-like quasi-particle) in Fermi liquids, Anderson proposed that there are two separate quasi-particles, one for charge (the 'holon') and one for spin (the 'spinon') and that an added electron 'falls apart' into these two objects! More precisely, the electrons in the system have re-organised themselves so that charge and spin disturbances can become widely separated in space.

There is evidence of this spin–charge separation in the one-dimensional analogue of the cuprates. Consider a chain with one electron per site on the chain. As in the cuprates, electrostatic repulsion makes this an antiferromagnetic Mott insulator. So, the direction of the spin on each electron is antiparallel to that of its neighbours. Suppose that we remove a spin-down electron from this chain (Figure 2.6(a)). Now, we can move the electron on the left of the vacancy onto this site, i.e. the charge disturbance (holon) moves to the left. Note that we have created a defect in the antiferromagnetic spin alignment – there are two neighbouring electrons with *parallel* spins. This is a spin disturbance (spinon) relative to the original state. If we iterate this procedure with the new vacancy, we can move the charge disturbance (holon) far away from the defect in the spin arrangement (spinon) (see Figure 2.6(b)).

This illustration of 'spin–charge separation' can be formalised into a quantitative theory in one dimension. This has led to the search for quasi-one-dimensional compounds that might exhibit this behaviour. Strontium copper oxide ($SrCuO_2$) is a promising candidate with copper–oxygen chains (instead of planes). It is a Mott insulator in its pristine state and an experiment like that suggested in Figure 2.6 can be done by ejecting electrons with light. When this is done there is evidence for seeing two moving particles instead of the single empty site left behind. Although the two-dimensional problem relevant to the cuprate superconductors has no mathematical solution so far, our progress with their one-dimensional relatives is encouraging.

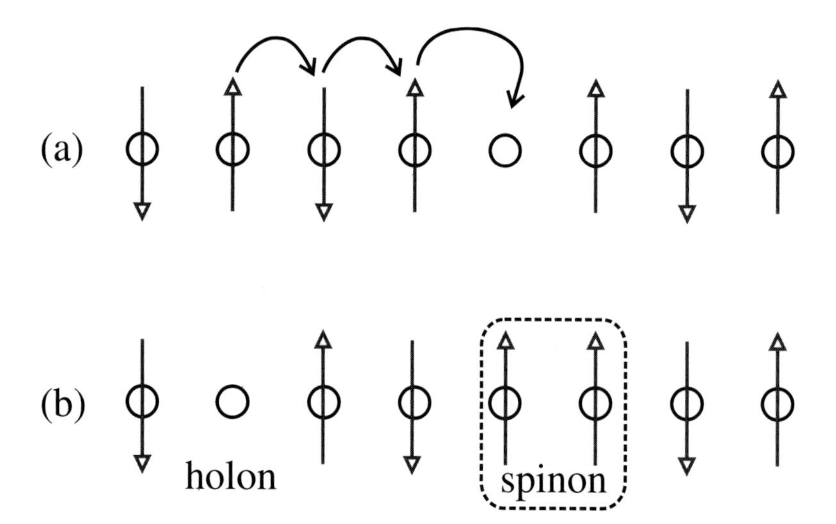

Figure 2.6. In one dimension, the electron can appear to fall apart into separate entities carrying the electrons' spin (the spinon) and charge (the holon). (a) By removing electrons from an ordered antiferromagnetic state, we remove both some spin and some charge. As electrons hop through the lattice we reach a state such that (b) the disruption in the spin ordering has moved away from the disruption in the charge ordering. This gives a crude picture of spin–charge separation. Philip Anderson has made the bold suggestion that this physics of one dimension may apply also to the high-temperature cuprates, in which the copper–oxide planes are effectively two-dimensional. Considerable theoretical and experimental effort is being invested in exploring this possibility.

2.5.2 Fractional charges

Other condensed-matter systems also exhibit novel coupling between spin and charge whereby, in contrast to the electron, other proportions of electrical charge and magnetism are bound together in the elementary excitations. An example can be found in semiconductors in high magnetic fields.

Semiconductor devices, such as the field-effect transistor, make use of electrons confined at the interface between two semiconductors. This is an effectively two-dimensional sheet so that reduced spatial dimension again comes into play. Suppose that a current, I, is flowing down the sheet due to an applied voltage from a battery. In addition, the sheet is placed between the poles of a magnet so that the magnetic field is perpendicular to the plane of the sheet of electrons. The flow of electrons bends round due to the Lorentz force from the magnetic field. This generates a sideways voltage, $V_{H,}$ *perpendicular* to the current flow. The Hall voltage and the current are proportional to each other, so we can define a Hall resistance: $R_H = V_H / I$.

For small magnetic fields, the Hall resistance increases smoothly with the magnetic field. For strong fields, its evolution develops steps and plateaux. In fact, it becomes *quantised* at particular fractions of h/e^2: i.e. $R_H = h/(\nu e^2)$ for $\nu = 1, 2, 3, \ldots$. This is the 'integer quantum-Hall effect' ('integer' because ν is a whole number). This quantisation is so accurate and robust against extraneous effects such as disorder that it has become the metrological standard for electrical resistance.

For very clean systems and even lower temperatures, we observe a 'fractional quantum-Hall effect' as well, with additional plateaux of the Hall resistance corresponding to fractional values of ν, such as $\nu = 2/3$, $2/5$, $1/3$ (see Figure 2.7). The discovery and exposition of the fractional effect won Laughlin, Störmer and Tsui the 1998 Nobel Prize. The most spectacular property of this system is that charge excitations come in fractional units of the elementary charge. When the Hall resistance of the system is at a plateau value of $3h/e^2$, the quasi-particle appears to carry a charge of $e/3$!

Have we split the electron? No. This fractional charge emerges from this electron quantum fluid through the co-operation of many electrons. In fact, all the electrons have conspired to organise themselves into a novel state. Here, the electronic motion is highly correlated to electrons circulating around each other in small groups of three, continuously changing from one group to the next. From this dance of the electrons emerges the physics of fractional charge.

Interestingly, the idea of magnetic flux quanta (in units of h/e) from superconductivity theory enters this story as well. For the $\nu = 1/3$ plateau to occur, the magnetic flux density has to be near three flux quanta per electron. It turns out that an elementary excitation of this system is the introduction of a unit flux quantum which, from the flux-per-electron counting, corresponds only to one third of the electronic charge. Again, it should be emphasised that these fractional charges are not theoretical constructs. Indeed, they have been observed directly in recent experiments.

2.5.3 Skyrmions

The integer quantum-Hall state also has interesting quasi-particles. In the previous section, we have assumed that all the electron spins are aligned

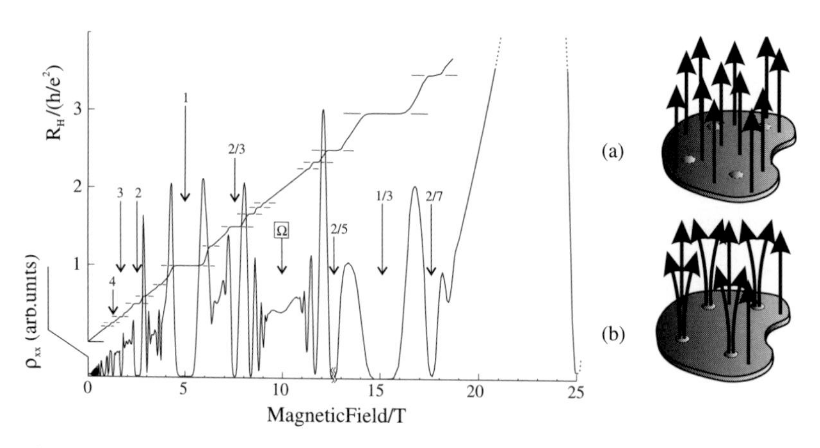

Figure 2.7. The fractional quantum-Hall effect is an example of how fractionally charged objects can appear in a condensed-matter system. Electrons confined in two dimensions exhibit unusual stability when the number of electrons per magnetic flux (the *filling*) is a whole number. This 'integer quantum-Hall effect' manifests itself as dips in the resistivity, ρ_{xx}, or plateaux in the Hall resistivity, R_H. The unexpected observation of dips (and corresponding plateaux) at fractional fillings was first understood by Bob Laughlin, who showed how fractionally charged objects must be responsible. Jainendra Jain showed how the fractional effect is related to the integer effect, by (a) electrons binding to magnetic flux to form (b) composite particles, leaving one free magnetic flux per particle. In other words, the fractional effect is the integer effect for the composite particles.

with the external magnetic field. This is the Zeeman effect. Recent experiments have shown that applying pressure to the solid can weaken the Zeeman effect, so that spin reversals are now possible.

It turns out that, even in the absence of an aligning field, the electrons in the quantum Hall regime prefer to align their spins parallel to each other. In other words, it is an intrinsic ferromagnet, like an iron bar. Consider adding an electron to this ferromagnet. In the integer Hall state, all the quantum states of one spin state have been occupied and the Pauli exclusion principle forces the additional electron to occupy a state with its spin antiparallel to those of its neighbours. The inherent ferromagnetism of this electron gas does not favour this misalignment and so the neighbouring electrons attempt to rotate their spins to smooth out the disturbance. This may involve up to twenty spins around the new electron (Figure 2.8). We see that adding one electron to the quantum-Hall system

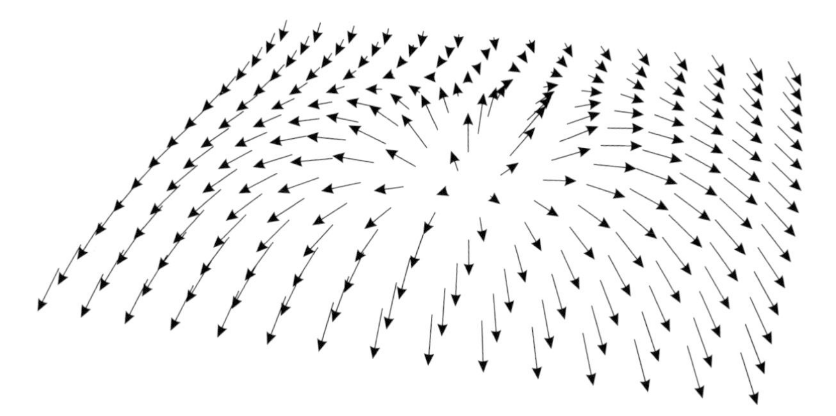

Figure 2.8. A skyrmion is a localised spin arrangement (known as a texture). Close to the centre of the skyrmion the electron spins align in one direction but far away this direction is reversed. This spin particle has been discovered in quantum-Hall systems.

results in the addition of one unit of charge *e* but *many* units of spin. This large-spin single-charge quasi-particle is called a *skyrmion,* named after Skyrme who originally discussed such objects in the context of topological field theories in high-energy physics. This skyrmion has finally been observed in the completely different arena of condensed-matter physics.

2.6 Speculations

One of the attractions of condensed-matter physics is that theoretical progress has implications for applications as well. Our understanding of electrons in solids, which is based on Landau's quasi-particle picture, underpins the silicon-based semiconductor technology that has transformed our lives in this century.

The discovery of high-temperature superconductors in 1986 by Bednorz and Müller has already opened up new opportunities in superconducting technology, which previously had been ruled out because of the expense and inconvenience of liquid-helium cryogenics. The absence of any electrical resistance allows the storage and transmission of electricity with no losses. The manufacture of superconducting cables is becoming a commercial possibility. Large magnets using superconducting coils are in use today in a wide range of applications, from particle accelerators to medical magnetic resonance imaging. The Meissner effect (expulsion of magnetic flux) can also be exploited to achieve magnetic levitation, leading to the prospect of friction-free trains with superconducting coils gliding over magnetic tracks.

On a smaller scale, superconductor electronics are also in sight. The cuprate superconductors are already in use in superconducting quantum interference devices (SQUIDs) which can detect minute changes in magnetic fields. These devices have a wide range of applications, from medical instruments to non-invasive testing for fractures in aircraft wheels.

The interplay of magnetism and electrical conduction has also recently found application in magnetic storage devices such as computer hard discs. Here one uses materials whose electrical resistance is strongly affected by a magnetic field (namely materials exhibiting 'giant magneto-resistance'). There are probably many such possibilities using strongly correlated materials in which, as we have seen, the competition between magnetism and electrical conduction can play an important role. It is, for example, interesting to note that manganate relatives of the cuprates have even higher magneto-resistances (so-called 'colossal magneto-resistance').

Looking further into the future, one might predict that a deeper understanding of the new types of exotic excitation arising in strongly correlated materials may lead to technologies every bit as rich as the single-electron quasi-particle physics which dominates the semiconductor device industry at present. Indeed, if technological dreams such as quantum computers are to be realised in practice, we will need to exploit the macroscopic consequences of quantum mechanics which routinely dominate the physics of materials of current theoretical interest.

Just as the turn of the last century saw the discovery of the electron and the key to understanding simple metals in terms of the electron quantum fluid, we are now seeing an increasing need for more exotic quantum fluids to account for the behaviour of more complex metallic states. Already we have seen materials in which the immutable electron should actually be viewed as separating into its magnetic and charged components. Also, fractionally charged objects have been identified in semiconductor devices and the dreams of particle theorists are being realised in condensed-matter systems.

These exciting developments are fuelled by the ever-increasing complexity of structures that material scientists can produce. From high-temperature superconductivity to colossal magneto-resistance, we are continually being surprised by the new phenomena that nature throws us as we investigate more complex systems. The search for new concepts to understand these phenomena is a major challenge facing the condensed matter physicist at the start of this new century. Technological applications are surely only a matter of time.

2.7 Further reading

Anderson, P. W. 1995 Condensed matter – the continuous revolution, *Physics World*, December.
Coleman, P. 1995 Condensed matter – correlated electron systems, *Physics World*, December.
Dung-Hai Lee and Shou-cheng Zhang, 1996 Electrons in flatland, *Scientific American*, March.
Rice, T. M. 1999 *Physics World*, December (and other articles in the issue).

3
Atom optics: matter and waves in harmony

William L. Power

Institute of Geological and Nuclear Sciences, Lower Hutt, New Zealand

3.1 Introduction

The science of optics has a long and distinguished history. It has led to many of the technologies which are important in our lives, such as the telescope and microscope, the laser and optical lithography, which is used mainly in the fabrication of microchips. Atom optics, which is the application of optical techniques to manipulate the wave properties of atoms, has the potential to be of great benefit to mankind too.

The laser has been at the heart of many of the most important scientific and technical innovations of this century. There can hardly be a home in the UK that does not possess one in the form of a compact-disc player or a CD-ROM. When we talk on the phone, or send messages on the Internet, there is a high probability that our message will be encoded by a laser into pulses of light to be passed through optical fibres. A laser produces 'coherent' light, which, roughly speaking, means that the photons which it emits travel in a synchronised way. In the last few years physicists have been discovering that it is possible to control the coherence properties of atoms and are now developing tools similar to those which take advantage of laser light. In the next decade we can look forward to seeing the equivalents of lasers, fibres, mirrors and waveguides for atoms. Already we have been able to perform interferometry experiments using beams of atoms and these have proved to be extremely sensitive measurement devices, allowing us to test physical theories and measure fundamental constants to a level of precision not previously possible.

Will these developing technologies ever be as ubiquitous as their optical forebears? In this article we discuss the anticipated uses of this technology both for fundamental science and for industrial applications. Starting from an explanation of the meaning of coherence and a comparison of the properties of photons and atoms, we describe the development of the science of atom optics. The tortuous route towards the great achievement of producing a Bose–Einstein condensate is described and we explain why this is the most important step towards a laser-like source of atoms. Numerous other optical elements for atoms have been built or are on the drawing board, so we describe how these may be used to manipulate atoms with a precision that has never before been achieved and look ahead to what we can learn about physics using these tools and at how they can be put to practical use.

It is only since the development of quantum mechanics earlier this century that we have become aware that atoms and other massive particles such as electrons and neutrons have wave-like properties. The duality between the wave and particle properties both of light and of atoms has an interesting history.

In the late seventeenth century two different models were proposed to describe the properties of light. Newton took the view that light was composed of a stream of 'corpuscles', whereas Huygens advocated a picture in which light is described by waves. For most of the eighteenth century Newton's corpuscular picture was the most widely accepted viewpoint, partly because of Newton's high esteem amongst his peers and partly because of doubts regarding the nature of the medium through which light waves would propagate. However, by the early years of the nineteenth century a number of interference and diffraction experiments had been performed by Young, Fresnel, Poisson and others, which could only be described in terms of waves. The wave description of light was then the clearly favoured theory, until Einstein's model of the photoelectric effect in 1905 re-introduced a particle-like interpretation. This provided a major step towards the development of quantum theory, which eventually led to the current description of light in terms of photons possessing both wave and particle properties.

Now let us compare this with the development of atom optics. The picture of matter as being composed of indivisible massive particles, called atoms, took hold in the late nineteenth century. Later de Broglie proposed that massive particles had an associated wave and this idea was developed by Schrödinger, leading to the famous equation which bears his name. Experimentally the wave properties of massive particles were first demonstrated by Davisson and Germer in 1927 in their electron-diffraction experiments. Subsequently Estermann and Stern demonstrated diffraction of helium atoms and hydrogen molecules from lithium fluoride crystals in 1930. Their experiments effectively started the field of atom optics, insofar as they were the first to demonstrate the wave-like properties of atoms.

In the last fifteen years there has been a renewal of activity in atom optics, culminating in the recent production of gaseous Bose–Einstein condensates (BECs) in which macroscopic numbers of atoms occupy the same quantum state. In such a condensate the waves associated with each of the atoms are in phase with one another in a way that is directly analogous to the behaviour of photons in a laser (Figure 3.1).

As the development of a laser-like source of atoms proceeds, physicists have built and continue to develop the equivalents of optical elements such as mirrors and waveguides with which to manipulate the resulting atom beams. With these new tools many exciting new experiments become possible. For example, the fact that atoms have mass has allowed the use of atom interferometry to probe the nature of gravity with greater sensitivity than ever before; indeed, atom interferometry experiments are often among the most precise in science. The complex internal structure of atoms has been used to gain new insight into the process of quantum measurement and the interactions of atoms with one another have revitalised areas of statistical physics, which have found new uses in describing the features of BECs. On a practical level, atom optics has the potential to allow us to manipulate and assemble atoms on a microscopic scale in ways previously impossible.

3.2 The physics background

Some of the most important characteristics of a light source are its coherence properties. It is the high degree of coherence in laser light that differentiates it from most other sources of light. Prior to the development of the laser virtually all experimental light sources were thermal, but now lasers are the preferred light source for many, probably most, optical experiments. In the same way, until very recently experiments in atom optics had to rely on thermal sources, such as beams produced by ovens, but now, with the development of laser-like sources of atoms, it is expected that these will in

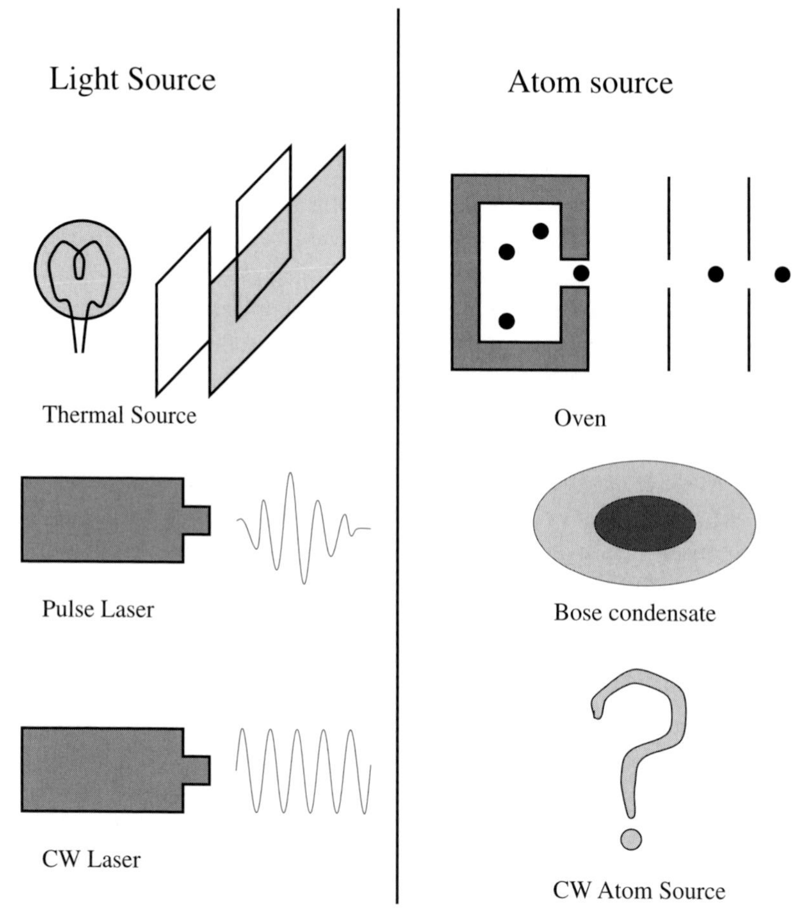

Figure 3.1. A comparison of the development of light sources and atom sources. Until the laser was developed in the 1960s, optical experiments were performed using thermal light sources and filters. Lasers were able to improve greatly on the coherence properties of thermal light sources. The first lasers produced pulses of light, but now most lasers produce a continuous beam of coherent light (CW lasers). In contrast, most atom-optics experiments still use thermal sources of atoms produced in an oven. Bose–Einstein-condensate experiments put atoms into a single quantum state; an individual condensate can be considered equivalent to a single laser pulse. Much current research is dedicated to producing a continuous coherent beam of atoms equivalent to the light from a CW laser.

future prove to be of considerable benefit to experimenters. However, what do we mean by coherence and why is it so important?

Roughly speaking, coherence is the property that describes how the phase of a source of waves fluctuates in time. In the case of a thermal light source, light is emitted at random times and the phase associated with the light waves varies randomly (except over very short time scales). In contrast, in laser light the phase of the light remains correlated over much larger intervals of time. Having been produced in a uniform manner, the photons, or wavepackets of light, 'march in step' rather than being out of sync with one another.

The coherence properties of a light source are often described in terms of a coherence length. This is the distance that light can travel in the time it takes for the phase of the light to become randomised. For a good, well-filtered, thermal source of light this may typically be a few centimetres, whereas for laser light it can be as large as hundreds of kilometres.

Photons of light have no rest mass and travel at the speed of light. The situation is different for massive particles. For a particle of mass m moving at velocity v the particle can be described by a matter wave with a wavelength $\lambda = h/(mv)$, where h is Planck's constant. This wavelength is often referred to as the de Broglie wavelength after the French physicist who first suggested it. In a filtered thermal beam of massive particles each individual particle may have a wavelength λ close to that of the other atoms in the beam, but the waves associated with the particles are out of phase with one another. However, in an atom laser the aim is to put all of the atoms into the same motional state so that the associated waves are in phase. In a gas of identical bosons (bosons are particles with a quantum spin that is an integer multiple of Planck's constant – this includes several kinds of atoms), this occurs when the de Broglie wavelength exceeds the average spacing between the atoms.

The behaviour of such a gas was first described by Bose and Einstein in the 1920s. They described how a gas cooled to such a low temperature that the de Broglie wavelength of the atoms was greater than their mean separation would 'condense' into a single quantum state. At the time this phase transition was regarded as being of purely academic interest since the temperatures required were far below those available to experimenters. According to Bose and Einstein, if a gas of bosons is cooled below a critical temperature T_c, the atoms all end up occupying the lowest-energy quantum state available.

3.3 Atom optics

Although the first experiments to demonstrate the wave properties of atoms were carried out in the 1930s, the field of atom optics remained relatively little investigated until the the late 1980s and early 1990s, when a number of groups started to develop experiments to demonstrate interference using atoms. Some of the first results to appear were the atom-interferometry experiments of Carnal and Mlynek in Konstanz and Pritchard and co-workers in MIT (both sets of results were published in 1991). The Konstanz experiment was an atomic version of Young's double-slit experiment in which helium atoms passed through a pair of microfabricated slits 8 μm apart in a thin gold foil and produced an interference pattern in a detector. The MIT experiment (Figure 3.2) used sodium atoms in a more complex set-up using three micro-fabricated gratings. Within a few months several other research groups had also demonstrated different configurations of atom interferometers.

Another important element in optical systems for atoms is the equiv-

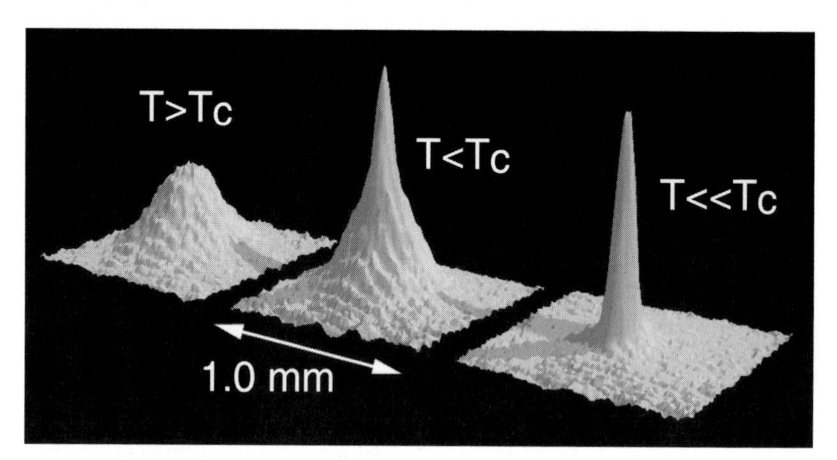

Figure 3.2. The experimental demonstration of Bose–Einstein condensation by Ketterle's group at MIT. The plot shows the density of atoms 6 ms after their release from an atom trap. The hotter atoms travel faster, so that, above the transition temperature, there is quite a broad distribution of atoms. At the transition temperature a sharp peak emerges, due to condensation of atoms into the trap ground state. Well below the transition temperature all of the atoms are in the ground state and only the central peak remains. (Figure courtesy of Wolfgang Ketterle.)

alent of a mirror for atoms. The most common approach for constructing an atomic mirror is to use evanescent waves. This works on the principle that the interaction of an intense quasi-resonant light field with the internal electronic structure of an atom creates a force on the atom. The result of this 'dipole force' is that atoms are attracted to regions of space where there is intense 'red-detuned' light, which has a frequency less than that of the internal atomic transition, and repelled from regions of intense 'blue-detuned' light, which has a higher frequency than the electronic transition of the atom.

So what is required to reflect an atom is a sharply defined region of intense blue-detuned light, which can be produced using the evanescent wave which is produced when light undergoes total internal reflection within a prism. The evanescent wave is a light field above the surface of the prism that decays exponentially with the height above the prism on a scale that is approximately that of the wavelength of the light. An atom falling towards such a prism will 'bounce'. The concept of an evanescent wave mirror was first proposed by Cook and Hill in 1982 and first demonstrated experimentally by a group of researchers in the USSR in 1988. Subsequently a group at the Ecole Normale Supérieure in Paris was able to observe several bounces by atoms on such a mirror, in effect creating an 'atomic trampoline'. It is also possible to use evanescent waves to line the interior of hollow optical fibres and in this way create a hose down which atoms can travel. In addition to those described here, an ever increasing range of coherent optical elements for atoms have now been made, or are being developed (see Arimondo and Bachor (1996) for further reading).

Perhaps the most important optical equivalent for atoms which is currently being developed is a laser-like source for atoms. Such a source would produce a highly coherent beam of atoms like that found in a BEC. Although the idea of a BEC has been around for a long time, the experimental realisation of a condensed gas has been the culmination of many years of hard work. During the 1980s much work was done to try to achieve laser cooling of atoms into a condensed state. In its basic form, laser cooling uses scattering of light at a frequency below that of an atomic resonance to slow fast-moving atoms selectively via the Doppler shift. More advanced laser-cooling techniques such as sub-Doppler laser cooling and laser 'sideband-cooling' techniques have subsequently been developed (see Chu and Wieman (1989) for a selection of articles on various aspects of laser cooling). Laser cooling brought the temperatures of trapped atoms down to the microkelvin range, but still the approach was not quite successful at producing the high densities and extremely low temperatures required to achieve a BEC. It was only later, when an additional process of evaporative cooling was used, that these conditions were satisfied. Evaporative cooling works according to the same principle as that by which a hot cup of coffee cools; the hottest atoms are allowed to leave a trapped cloud of atoms and those that remain are allowed to re-thermalise, which has the effect of reducing the temperature of those that remain. Members of a group led by Cornell and Wieman at the National Institute for Science and Technology in Boulder, Colorado, were the first to produce a condensate using this technique. They produced a condensate consisting of approximately 2000 rubidium atoms cooled to 100 billionths of a degree above absolute zero, the lowest temperature ever produced, making the atom cloud the coldest place in the solar system – a feat that was soon followed by Ketterle's group at MIT using sodium atoms.

Producing BECs is not the end of the story for atom lasers. When a single Bose condensate is released from its confining potential it behaves much like a single pulse of laser light, thus Ketterle and co-workers have been able to couple out a large series of pulses from a single condensate. Attempts to use such pulses to perform atom-optics experiments such as interferometry are being made, but ultimately it would be nice to produce a continuous source of coherent atoms analogous to a continuous-wave laser. At the time of writing, the production of coherent streams of atoms lasting for as long as 100 ms in experiments performed at the Max Planck Institute for Quantum Optics in Germany has just been announced. They achieved this by effectively punching a 'hole' in the magnetic trap used to confine a BEC, through which atoms are then able to leak continuously. It would still be desirable to make the whole process of trapping, cooling and condensing of atoms a continuous one. Schemes for the continuous loading of atoms into a degenerate gas have been considered by a large number of groups and experimental steps towards this end are under way. Some schemes, such as that of Mlynek's group at the University of Konstanz, have atoms continuously pumped into a standing wave optical potential using evanescent waves. Similar techniques, which use magnetic fields instead of evanescent waves, are being developed by Hinds' group at the University of Sussex.

3.4 Applications

At present two main applications are being implemented using atom optics. One is high precision atom interferometry; the other is fabrication of structures on the atomic scale.

Atom interferometry can offer a number of benefits over other forms of matter interferometry. Atoms are less susceptible to stray fields than are electrons and atom beams are easier to produce than are neutron sources. In addition they have a considerably higher mass, which is of significant benefit for the detection of gravitational phenomena. In particular, the experiments of Kasevich and Chu have been able to detect changes in the earth's gravitational field at an extremely sensitive level, detecting changes in the acceleration due to the earth's gravitational field down to about one billionth of g. With such a high accuracy, it has even been suggested that atom interferometry could be sensitive to quantum mechanical fluctuations in the gravitational field, offering a way to test quantum theories of gravity. Rotations can also be very sensitively measured using atom interferometers, as was demonstrated in a recent experiment by Gustavson, Bouyer and Kasevich in which an atomic gyroscope was created in an interferometer using caesium atoms. Maybe in future, if the technology can be miniaturised, atom interferometers could form the basis for very accurate navigational instruments.

A second important application for atom optics is in the fabrication of microstructures. With high-precision optics it may be possible to position atoms onto a substrate with great accuracy, which may allow structures to be built on a much smaller scale than current optical lithographic techniques allow, mainly due to the short de Broglie wavelengths of the atoms. Furthermore, many of the techinques for depositing atoms operate in parallel, with many identical microstructures created simultaneously adjacent to one another. This has clear benefits in terms of efficiency and reliability. Periodic microstructures, in particular, can be efficiently implemented in this way. This technology has the potential for application in fabricating new generations of microchips on virtually an atom-by-atom scale and it could also form the basis for very-high-density data-storage devices.

The suggestions above represent only a couple of the possible applications for atom optics. If the development of the technology proceeds at a similar pace to that of laser optics, we can expect that far more uses will be found than those so far envisaged.

3.5 Conclusions

Although the wave-like properties of atoms were first demonstrated in the 1930s, it is only in the last decade that atom optics has developed into a widely useful tool. The development of BECs has led to the possibility of laser-like sources of atoms, which will further increase the scope of what can be done with this branch of physics.

3.6 Further reading

For an extended version of this article, including comprehensive references, see Power, W. L. 2000 Atom optics: matter and waves in harmony, *Phil. Trans. Roy. Soc.* A, **358**, 127.

For a collection of technical articles on applications of coherent atom optics, see Arimondo, E. and Bachor, H. A. (editors) 1996 Special issue on atom optics. *Quantum Semiclassical Optics* **8**, 495.

For a selection of articles on laser-cooling, see Chu, S. and Wieman, C. 1989 Laser cooling and trapping of atoms, *J. Opt. Soc. Am.* B **6**, 2020.

4
Quantum electronics: beyond the transistor

Alexander Giles Davies

The Cavendish Laboratory, University of Cambridge, Madingley Road, Cambridge CB3 0HE, UK

4.1 Introduction

In the late 1930s, Mervin Kelly, the visionary Director of Research at Bell Telephone Laboratories in New Jersey, dreamed of eliminating the slow, bulky and unreliable vacuum tubes and electromagnetic switches upon which his telephone network relied. Thinking back to the compact 'crystal detectors' used at the beginning of the century in radio receivers, he imagined replacing vacuum tubes by small, robust, low-power devices with no moving parts, in which the switching took place electronically. Crystal detectors were used to rectify the oscillating current generated by radio waves into the direct current required for the headphones. These detectors were made of a piece of a suitable crystalline material such as galena (lead sulphide), copper oxide or silicon sandwiched between two metal electrodes. The crystalline substances central to these early electrical devices belonged to a class of materials called 'semiconductors' and, as the unique properties of these materials began to be revealed and understood, semiconductors increasingly began to make a profound impact on the world.

In 1945, Kelly assembled a team to perform fundamental research into solid-state physics, in order to find out more about semiconductor materials and to assess their potential for electrical devices. One member, Walter Brattain, had worked on copper oxide crystal rectifiers for many years and envisaged incorporating a third electrode (analogous to the 'grid' electrode in vacuum tubes) to make a solid-state switch or amplifier. The team also included theoretical physicist John Bardeen and was to be led by William Shockley (Figure 4.1(a)). Two years later, the Bell Labs team demonstrated a new electrical amplifier, which they called the 'transistor'. Unlike the prevailing vacuum tube amplifiers, the transistor was a solid-state device built from a piece of semiconductor crystal. Its invention sparked a revolution in electronics and communication technology that continues to rage unabated over fifty years later. However, one of the most striking aspects of the progress of semiconductor science over the last fifty years is how the commercially driven technological developments have occurred alongside advances in fundamental physics obtained from investigation of the same semiconductor devices. The basic building blocks of computer and communication technologies are perfect for the study of electrons and their interactions with each other and with their environment – the fundamental interactions of one of nature's most fundamental particles.

In this article, we will investigate the symbiotic relationship between the technological and the fundamental aspects of these electronic devices and review some recent fascinating highlights of semiconductor physics and technology. However, we will also look at a future generation of microelectronic devices in which the fusion of molecular biology, chemistry and physics will be producing breathtaking results. Let us start by examining semiconductors in more detail.

4.2 Semiconductors

The fundamental properties of semiconducting materials and the ways in which they differed from metals and insulators were only beginning to be appreciated at the time Bardeen, Brattain and Shockley got to work. One hundred years earlier, whilst performing his seminal investigations of electricity and magnetism at the Royal Institution, Michael Faraday identified a series of materials distinct from metals in that they conducted electricity poorly and possessed a strongly temperature-dependent conductivity that improved (rather than decreased) when they were heated. Although much theoretical understanding of electrical conduction in solids had been provided in the 1920s by Felix Bloch, Rudolf Peierls and Alan Wilson *inter alia*, semiconductor samples were polycrystalline and contained impurities that affected their properties unpredictably. Silicon and germanium were still thought by many to be metals. Furthermore, although metal–semiconductor junctions had been used for many years in

(a)

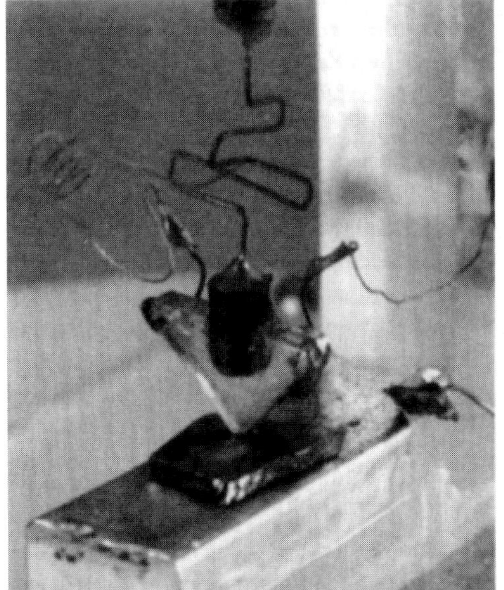

(b)

Figure 4.1. (a) From left to right, John Bardeen, William Shockley and Walter Brattain in 1948 after their invention of the transistor (courtesy of Lucent Technologies Bell Labs Innovations). (b) Walter Brattain's first bipolar point contact transistor. (Property of AT&T Archives. Reprinted with permission of AT&T.)

crystal detectors, they would not begin to be understood until the work of Walter Schottky and Nevill Mott in the late 1930s.

Bloch applied the theory of quantum mechanics to see what happened to electrons travelling through a crystal lattice. Quantum mechanics was the name given to the very successful new theory developed to explain phenomena at the atomic level. The mechanics introduced by Isaac Newton in the late seventeenth century that worked perfectly in explaining, for example, how the earth went around the sun, or the trajectory of a cannon ball, failed when it was applied to the motion of electrons. A central aspect of the new theory was that particles such as electrons could, under appropriate conditions, be better described as a wave (we will come back to this later). Bloch found that electrons with certain energies would not be able to propagate through the crystal because they would be diffracted by the periodic lattice structure. A series of energy bands separated by 'forbidden' energy gaps is formed, with the electrons constrained to have energies lying within the bands. Wilson developed this band theory further in 1931 and explained the distinction between metals, semiconductors and insulators in the following way (Figure 4.2(a)). The most energetic electrons in a metal partly fill a band (called the conduction band) up to an energy called the 'Fermi energy'. Under the influence of an electric field, these electrons acquire energy from the field and scatter into the adjacent empty states in the band above the Fermi energy. Their ability to respond to the field in this way results in an electrical current. In an insulator, however, the most energetic electrons lie at the very top of an energy band (the valence band). There are no empty states close in energy to scatter into because of the proximity of the 'forbidden' band gap, so insulators cannot conduct. However, the size of this band gap separating the filled valence band from the empty conduction band is crucial. At low temperatures, pure semiconductors insulate since all electrons fit snugly in the valence band. However, the band gap is sufficiently small that, at ordinary temperatures, some electrons are thermally excited into the conduction band. This results in a certain degree of conduction. Two important consequences of band theory should be noted.

First, the electrical properties of semiconductors can be tailored by incorporation of extrinsic impurities called 'dopants'. In fact, once the density of dopants exceeds a certain threshold, the semiconductor undergoes a transition to a metal, but one with a much lower concentration of conducting electrons than that in elemental metals such as copper and

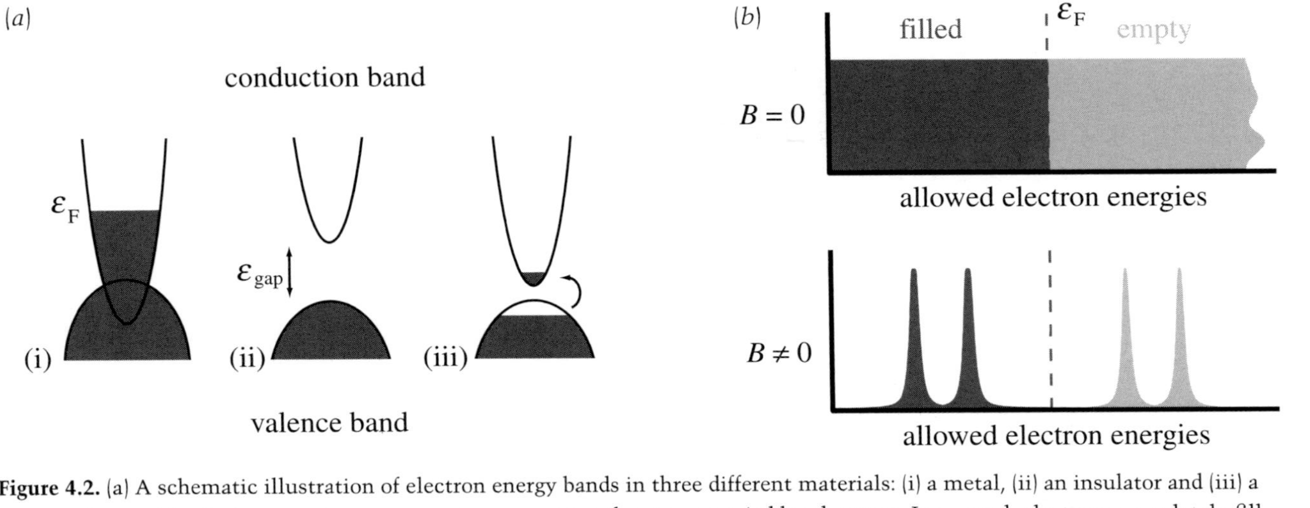

Figure 4.2. (a) A schematic illustration of electron energy bands in three different materials: (i) a metal, (ii) an insulator and (iii) a semiconductor. The dark grey regions represent energy states that are occupied by electrons. In a metal, electrons completely fill the valence band and partly fill the conduction band up to the Fermi energy (ε_F). Under the influence of an electric field, the electrons at the Fermi energy acquire energy from the field and scatter into the adjacent empty states in the band. Their ability to respond to the field in this way results in an electrical current. In an insulator, electrons completely fill the valence band, which is separated from the empty conduction band by the 'band gap' (ε_{gap}). There are no empty energy states close by for the Fermi electrons to scatter into because of the proximity of the 'forbidden' band gap, so insulators cannot conduct. A semiconductor is similar to an insulator but with a much smaller band gap, allowing some electrons to be thermally excited into the conduction band at room temperature. (b) Schematic diagrams showing the energies that electrons in a two-dimensional system are allowed to have at zero magnetic field ($B=0$) and at finite perpendicular magnetic field ($B \neq 0$). The magnetic field splits the continuum of allowed energy states into a ladder of discrete levels.

gold. Second, band theory gave an explanation of the 'hole' conduction observed in some semiconducting materials. This is when the current appears to be carried by *positively* charged particles called 'holes', rather than by the usual negatively charged electrons. This situation occurs in materials with a nearly full valence band. A good analogy is that, just as a few drops of water can trickle along an otherwise empty tube (electron conduction), air bubbles in a tube nearly full of water can also move (hole conduction). Conduction by this anomalous particle had been identified in studies of the 'Hall effect', discovered in 1879 by Edwin Hall, who found that the current in a thin film of gold was deflected by a perpendicular magnetic field, resulting in a transverse voltage (Figure 4.3(a)). Although the polarity of the Hall voltage is consistent with conduction by negative electrons in most materials, results of some studies suggest that conduction by positive particles is occurring. The Hall effect became a useful means for determining the type (electrons – n-type; holes – p-type) and concentration of particles in semiconductors, but was to have a glorious future. As we shall see, one hundred years later, it would be central to two Nobel Prizes for Physics.

4.3 The field effect

In 1945, Shockley started working on a mechanism now known as the 'field effect'. The aim was to modulate the current in a thin film of silicon by an electric field produced by a surface metal plate (the 'gate'). As a concept, the field effect extends back to the beginning of the century. Nevill Mott recounts in his autobiography how his father Charles attempted to observe the effect with J. J. Thomson at the Cavendish Laboratory between 1902 and 1904, shortly after Thomson had discovered the electron in 1897. This experiment failed since they tried to modulate the current in a metal (rather than a semiconductor) in which the electron concentration is too high for the effect to be observable. Shockley and his collaborators were unaware of the pedigree of the field effect until they tried patenting the idea and discovered that Julius Lilienfeld had preempted them with three patents filed in the late 1920s for field-effect semiconductor devices. Although it is not known whether Lilienfeld built any of his devices, his structures are remarkably prescient of successful devices fabricated over twenty years later.

Shockley's failure to observe the field effect led Bardeen to propose that

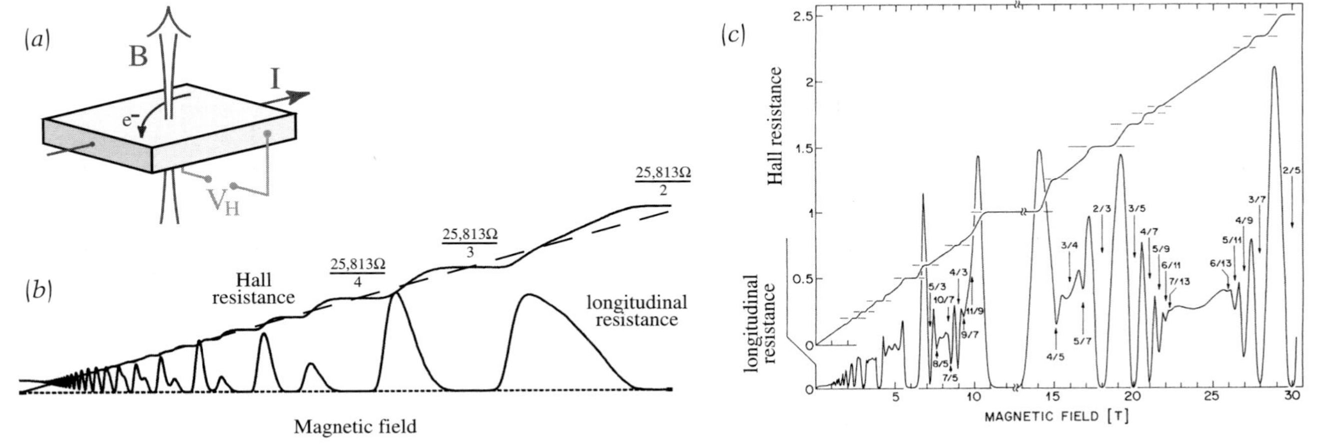

Figure 4.3. The Hall effects. (a) A schematic illustration of how a current of electrons (e^-) is deflected by a perpendicular magnetic field B to produce a Hall voltage V_H. The Hall resistance is obtained by dividing the Hall voltage by the magnitude of the electron current, I. (b) A schematic illustration of the integer quantum Hall effect. The Hall resistance increases with the magnetic field in a step-wise fashion, jumping up from one plateau to the next (the dashed line shows the classical Hall effect – a linear increase with the magnetic field), whilst the longitudinal resistance oscillates up and down. The value of the Hall resistance at each plateau is given by $h/(ie^2)\ \Omega = 25\ 813/i\ \Omega$ (where i is an integer that is different for each plateau) to a very high accuracy, better than an astonishing one part in one hundred million. (c) An overview of the rich spectrum of integer and fractional quantum Hall effect features observed in a GaAs–AlGaAs heterojunction. The numbers labelling the features refer to the corresponding value of i in the formula $h/(ie^2)\ \Omega$. (From Willett *et al.* 1987 *Phys. Rev. Lett.* **59**, 1776–9.)

electron traps ('surface states') form at semiconductor surfaces and that these traps prevent the gate field penetrating. In 1947, Brattain and co-worker Robert Gibney discovered that an electrolyte such as water between the semiconductor and the gate neutralised the surface states. Brattain and Bardeen found that they could just modulate the current flow from a tungsten point contact into silicon (and later germanium) by using a drop of water both to neutralise the surface states and, via a second electrode, to act as a gate. However, they were concerned about the frequency response. The device had to operate at audible frequencies for telecommunications, but the sluggish response of the electrolyte restricted operation to a few cycles per second. They replaced the electrolyte by a thin layer of germanium oxide, but this washed off and, with both electrodes now pushed into the surface, they discovered a new effect and invented the first transistor – the *bipolar point-contact transistor*. (See Riordan and Hoddeson (1997) or Seitz and Einspruch (1998) for excellent historical and scientific reviews of this period.)

4.4 Two transistors

Brattain and Bardeen found that the current between one electrode (the 'collector') and the germanium slab (the 'base') could be controlled by the potential on the second electrode (the 'emitter'). A small emitter–collector current not only resulted in a larger collector–base current but also modulated it. After further work, the bipolar point-contact 'transistor' (named by Bell Labs colleague John Pierce, who is also credited with the aphorism 'Nature abhors the vacuum tube') was operating at 1 kHz with a power gain of several hundred per cent; Figure 4.1(b) shows a demonstration transistor Brattain constructed in December 1947.

Shockley, feeling rather left out, worked feverishly on a device in which the emitter and collector electrodes would be consolidated inside the semiconductor, eliminating the clumsy and electrically noisy point contacts. He proposed a sandwich structure comprising a p-type region (the base) encased by two n-type regions (the collector and emitter) (Figure 4.4(a)). Small changes in the base bias led to exponential changes in the emitter–collector current analogous to a dammed river in which a small variation in the height of the dam produces a large change in water flow. The physical chemist Gordon Teal realized that more reproducible behaviour would be obtained if the detrimental scattering of electrons from grain

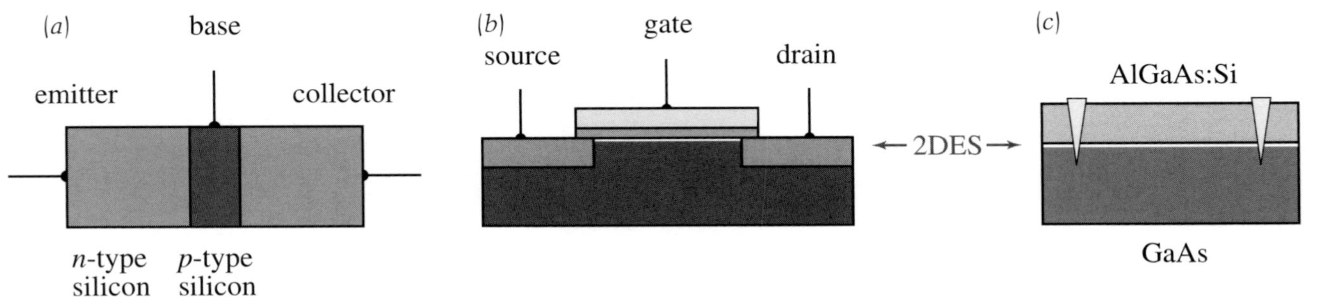

Figure 4.4. Schematic illustrations of (a) the bipolar junction transistor, (b) the silicon MOSFET and (c) the GaAs–AlGaAs heterojunction. A two-dimensional electron system is formed in the MOSFET and the heterojunction, indicated here by the white lines.

boundaries were eliminated. Together with Morgan Sparks, he developed a technique for pulling single-crystal germanium directly from the melt. They could change the doping between *n*-type and *p*-type by adding small amounts of an appropriate dopant element such as gallium or antimony to the melt, and in 1950, successfully fabricated Shockley's *bipolar junction transistor*. Shockley, Bardeen and Brattain ultimately received the Nobel Prize for Physics for their 'investigations on semiconductors and the discovery of the transistor effect' in 1956.

4.5 Transistor computing

Shockley immediately appreciated the transistor's potential for computing. In a 1947 interview he pointed out that 'For applications of this sort there are difficulties in applying vacuum tubes because of their size and the heat that they produce. It seems to me that in these robot brains the transistor is the ideal nerve cell.'

The desire to build a 'robot brain' has entertained engineers for several hundred years (see Shurkin (1996) for a thorough history of computing). During the seventeenth and eighteenth centuries, several (often untrustworthy) machines were built to perform simple arithmetic. The story of Charles Babbage's two unrealised visions – his Difference Engine (Figure 4.5(a)), a machine designed to calculate mathematical tables in a preordained manner, and his Analytical Engine, a programmable machine remarkably prescient of modern computers, which was designed to perform calculations according to instructions entered on punched cards – is well known. Analogue machines built around winches and pulleys gave way to more reliable electromechanical operation in 1944 with the IBM Automatic Sequence Controlled Calculator (the Mark I), designed by Howard Aiken. Researchers had also turned to vacuum tubes to increase the switching speed, albeit reluctantly. The COLLOSSUS, built in Britain's Bletchley Park in 1943 by a team including Alan Turing, was one of the first electronic computers. It employed 1800 vacuum tubes and was used exclusively for wartime cipher decryption. The first electronic digital computer was built at the University of Pennsylvania in 1946. The Electronic Numerical Integrator and Computer (ENIAC) used over 17000 tubes, weighed thirty tons, and consumed nearly 200 kW. ENIAC could add 5000 numbers per second and was used to calculate artillery-shell trajectories. It has been remarked that the energy required to calculate the trajectory of a shell was comparable to that of the explosive discharge needed to fire the shell itself! When the US Army decommissioned ENIAC nine years later because of the expense of operating it, it was still the world's fastest computer. Figure 4.5(b) shows the first public demonstration of the Automatic Computing Engine (ACE) in 1950, one of Britain's earliest general-purpose stored-program computers. It was based on a design by Alan Turing and built at the National Physical Laboratory; it used 800 tubes and could add 15 000 numbers in one second.

Although the transistors of the early 1950s switched more slowly than did vacuum tubes, it was clear that the inherent disadvantages of the latter (bulkiness, high power consumption, the requirement for continuous power, warm-up time, large amount of heat produced and frequent failure) would lead to their demise. The first purely solid-state digital computer was the Transistorized Digital Computer (TRIDAC) built by Bell Labs in 1954 for the US Air Force. It employed 700 point-contact transistors and rivalled digital vacuum tube computers such as ENIAC in computational speed. Figure 4.5(c) shows an early British transistor computer, the Elliott 803, of 1963.

In 1955 Shockley left Bell Labs to set up his own company in the San Francisco Bay area where he had once lived. Shockley Semiconductor Laboratory opened in February 1956, seeding the growth of high-technology companies in the area of California now known as Silicon Valley. Although Shockley recruited an excellent team, his company was not a success and many of his personnel resigned the following year to form Fairchild Semiconductor nearby, under the leadership of Robert Noyce and Gordon Moore.

4.6 Silicon, silicon dioxide, the integrated circuit and the microprocessor

Although Texas Instruments began as a geophysical company, in 1952 Vice-President Pat Haggerty decided that transistors were the future. Gordon Teal joined Texas from Bell Labs at the end of 1952 and devoted his expertise in crystal growth to the fabrication of single-crystal silicon. By 1954 Teal and his team had produced the first silicon bipolar junction transistor (Figure 4.4(a)). Silicon is more reactive than germanium and was more difficult to work with, but it has a larger band gap and so its electrical properties are less sensitive to temperature. Germanium transistors fail

(*a*)

(*b*)

(c)

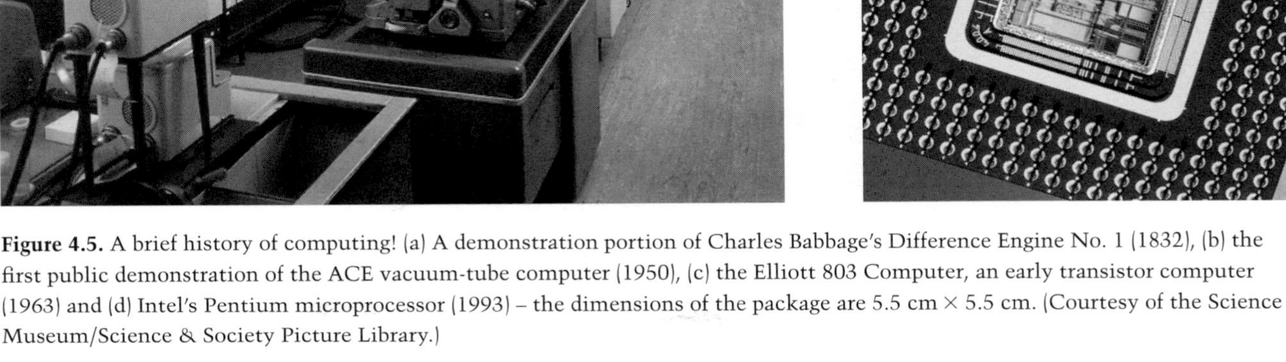

(d)

Figure 4.5. A brief history of computing! (a) A demonstration portion of Charles Babbage's Difference Engine No. 1 (1832), (b) the first public demonstration of the ACE vacuum-tube computer (1950), (c) the Elliott 803 Computer, an early transistor computer (1963) and (d) Intel's Pentium microprocessor (1993) – the dimensions of the package are 5.5 cm × 5.5 cm. (Courtesy of the Science Museum/Science & Society Picture Library.)

if they are heated to 70 °C, making them insufficiently robust for military applications, for example.

By the end of the 1950s, tens of millions of transistors were being produced each year and finding applications in fields as diverse as office equipment and satellites. However, as circuits became more complicated, they were increasingly difficult to assemble since each discrete component had to be individually wired to the next. The possibility of consolidating the components into a single integrated structure started to crystallise with a series of key developments that have engendered the massive integration and miniaturisation of high-speed switching, logic and memory circuitry over the last forty years.

The breakthrough came in 1958. Jack Kilby at Texas realised that, if conventional circuit elements such as resistors, diodes and capacitors were made from silicon, they could be incorporated with transistors on a single silicon substrate. As well as miniaturising circuits by consolidating the circuit elements and doing away with the interconnecting wires, this procedure would eliminate assembly errors. Kilby used photographic techniques to pattern the silicon wafer, introducing precise concentrations of dopants to specific areas and depths by the thermal diffusion of appropriate metals (a technique demonstrated in 1951 by John Saby at General Electric, which allowed the fabrication of the diffused-base bipolar junction transistor). However, at the same time, Robert Noyce at Fairchild was also considering the problem of interconnection, realising the absurdity of separating individual transistors fabricated on a silicon wafer only to reassemble them subsequently with soldered wires. In 1959, his colleague Jean Hoerni proposed coating the silicon surface with a thin layer of silicon dioxide (SiO_2) that, unlike Brattain's layer of germanium oxide, provided an insoluble, insulating protective sheet. This allowed the photographic patterning of fine interconnecting wires on the SiO_2 surface, with holes etched through the SiO_2 to allow access to the transistors beneath.

However, SiO_2 was discovered to have a further significant property – it passivates the surface states of silicon, allowing the electric field from a gate electrode on the SiO_2 surface to penetrate the silicon channel below. So, at Bell Labs in 1960, the field-effect transistor (the MOSFET – metal-oxide–semiconductor field-effect transistor, Figure 4.4(b)) was finally produced. This technology would prove cheaper and capable of higher device packing densities than its bipolar junction counterpart and hence the complementary MOSFET, developed in the late 1960s, which comprises an

n-type and a p-type MOSFET in series, has progressively replaced the junction transistor in integrated circuits. Since the complementary MOSFET draws power only when it is switching, it has led to the high component-packing densities found in present-day circuits which would otherwise be prevented by the devices overheating (but even a modern 'room-temperature' microprocessor operates above 100 °C, heated by its internal power dissipation!).

The early 1970s heralded two further developments – semiconductor memory and the microprocessor. In 1968, Noyce and Moore left Fairchild to found Intel and started making semiconductor memory circuits. Although the magnetic data-storage systems that replaced punched-card storage in the 1950s are still used today for archival purposes, the one-kilobit-capacity memory chips of the early 1970s pioneered cheap compact storage of vast quantities of information. A modern 256-megabyte complementary MOSFET dynamic random-access memory chip may contain several hundred million transistors with sub-micrometre-sized features packed in a postage-stamp-sized area. The microprocessor combines key computer circuitry in one versatile programmable chip. Intel's first 4004 microprocessor in 1971 contained 2300 transistors with features as small as 10 μm and a clock speed of 108 kHz. The Intel Pentium, shown in Figure 4.5(d), launched in 1993 had 3.1 million transistors with 0.8 μm features and a clock speed of 60 MHz. The Pentium III processor, launched January 2000, had 28 million transistors with 0.18 μm features and operated at 733 MHz. This processor has been superseded just as rapidly with >1 GHz models routinely available at the time of writing this chapter (January 2001). No doubt, even this specification will seem dated when this book is published! Microprocessors are the basis of modern technology, central to an extraordinary variety of medical, scientific, business, and consumer equipment. Jack Kilby was awarded the Nobel Prize for Physics in 2000 for his role in the development of the integrated circuit.

4.7 The two-dimensional electron system

It is not just the semiconductor microelectronics industry that has developed at an astronomical rate over the last fifty years. The fundamental solid-state physics research that originally led to the invention of the transistor has advanced just as rapidly, keeping pace with the technological developments and allowing a cross-fertilisation of ideas and technology between the pure and applied areas of the field. The versatility of semiconductor technology lies in its ability to create new designer materials and structures, that are not found in nature, the optical and electronic properties of which can easily be tailored. This makes semiconductor systems ideal 'laboratories' for basic research since we can create a specific material or structure to address a particular physical problem.

We will have a look at a few prominent themes from a vast literature (more details can be found in Davies (2000)). The most famous discoveries are perhaps the integer and fractional quantum Hall effects found in the two-dimensional electronic systems inherent to silicon MOSFETs and gallium arsenide heterostructures (described below). In addition, an exciting new field of nanostructure physics has emerged, in which the subsequent electrostatic confinement of these two-dimensional systems into one-dimensional wires, one-dimensional rings and zero-dimensional boxes, for example, has allowed the investigation of quantum mechanical electronic transport through nanometre-scale (1 nm is one millionth of a millimetre) architectures of ever-increasing complexity.

A silicon MOSFET is shown schematically in Figure 4.4(b). The *n*-type electrodes form rectifying contacts to the *p*-type substrate and so no current flows in the absence of an appropriate positive bias on the gate electrode. Such a bias sucks electrons into the channel between the electrodes and establishes a thin conducting layer of electrons at the semiconductor–oxide interface. However, it was realised in the late 1950s that these electrons might behave unusually since they are constrained to a layer less than one hundred-thousandth of a millimetre thick. What is so special about this? Well, this thickness is comparable to the electrons' quantum mechanical wavelength, with the consequence that they can form a two-dimensional electron system with the freedom to move only on a plane parallel to the interface. Measurements of these systems can reveal fascinating quantum mechanical phenomena since the restricted dimensionality forces the electrons to interact strongly with each other.

The application of a large magnetic field to a two-dimensional electron system has a particularly profound effect. We have already seen how a small perpendicular magnetic field deflects an electron current (the Hall effect), but, as the magnetic field becomes stronger, the electrons are increasingly forced to bend into circular trajectories (called cyclotron orbits) with progressively smaller radii. However, in intense magnetic fields, the rules of quantum mechanics come into play once again and restrict these cyclotron orbits to certain discrete sizes, each of which has a different energy. The magnetic field is said to 'quantise' the motion of electrons in the plane and splits the continuum of allowed electron energies into a ladder of discrete energy levels known as Landau levels (Figure 4.2(b)). This quantisation is manifested by the oscillatory behaviour of a number of physical properties as a function of the magnetic field, including the magnetic susceptibility (the de Haas–van Alphen effect), thermal conductivity and electrical conductivity (the Shubnikov–de Haas effect). The number of Landau levels occupied at any time depends upon the strength of the magnetic field – as the field is increased gradually, the number of filled Landau levels steps down abruptly.

4.8 The quantum Hall effect

The most famous discovery made in studies of two-dimensional electron systems is the integer quantum Hall effect which was discovered in 1980 and the subject of the 1985 Nobel Prize for Physics. In low-temperature measurements of the Hall effect in silicon MOSFETs, several researchers in the late 1970s found that the Hall resistance did not increase linearly with the magnetic field as expected, but instead small kinks and plateaux were observed. Subsequent experiments by Klaus von Klitzing, Gerhard Dorda and Michael Pepper showed that these plateau regions in the Hall resistance could be significant and they found that the Hall resistance in fact increases with the magnetic field in a step-wise fashion, jumping up from one plateau to the next (Figure 4.3(b)). The longitudinal resistance of the sample was found to oscillate up and down, with minima (which sometimes reached zero resistance) occurring at the same magnetic fields as the Hall plateaux and maxima occurring whenever the Hall resistance jumped from one plateau to the next. The value of the Hall resistance at each plateau was given by the formula $25\,813/i\ \Omega$ (where i is an integer that is different for each plateau) to a very high accuracy, better than an astonishing one part in one hundred million. (In fact, this formula is more correctly expressed as $h/(ie^2)\ \Omega$, where h and e are fundamental quantities of nature, namely Planck's constant, $h = 6.6\times10^{-34}$ J s, and the electron charge, $e = 1.6\times10^{-19}$ C.) It is worth thinking about the implications of this amazing experiment a little longer. The resistance of most materials is critically dependent on the size of the sample tested. However, if you take a silicon MOSFET of reasonable quality, irrespective of its size and shape, irrespective of whether it has chunks cut out of it or ragged edges, its Hall resistance will be given precisely by an integer fraction of h/e^2 Ω.

The reason this phenomenon is so resilient is that, in intense magnetic fields, all the conduction takes place along the edges of the sample. Electrons in the bulk of a sample simply execute circular cyclotron orbits and do not go anywhere, but those close to the edge repeatedly strike the edge and bounce along in conducting channels called 'edge states'. The number of edge-state conducting channels is equal to the number of occupied Landau levels and so depends upon the strength of the magnetic field. These edge states conduct in one direction only and electrons with forward and reverse momenta are physically separated on opposite sides of the device. For an electron to be scattered backwards it has to cross the entire device, so normal scattering of electrons by impurities does not affect conduction – this is why the longitudinal resistance is generally vanishingly small. Furthermore, it turns out that edge states are ideal one-dimensional conductors (see below), which have the quantum mechanical property of each contributing conductance e^2/h to the Hall conductance. Therefore, if there are i occupied Landau levels, there will be i conducting edge states, the Hall conductance will be ie^2/h and the Hall resistance will be the reciprocal of this, $h/(ie^2)$. This picture breaks down when the magnetic field is such that the number of occupied Landau levels is changing; this causes the Hall resistance to step up or down and makes the longitudinal resistance temporarily peak. This phenomenon, called the integer quantum Hall effect, is now used as the international resistance standard and has generated a massive international research effort.

4.9 Layered semiconductor devices

Over the last thirty years, techniques for fabricating a number of high-purity, compound semiconductor crystals such as gallium arsenide (GaAs) and aluminium gallium arsenide (AlGaAs) as well as layered devices called heterostructures have been developed. These comprise a series of different semiconductor materials grown sequentially one on top of the other, with the crystal lattice maintained throughout (epitaxial growth). Often the layers are only a few atoms thick. The optical and electronic properties of the constituent semiconductors can be combined to tailor new structures with new properties. Molecular beam epitaxy is perhaps the best-known growth technique; it is essentially a sophisticated form of high-vacuum evaporation that allows the fabrication of near-perfect crystals with extremely abrupt changes in composition and doping.

The GaAs–AlGaAs heterojunction (a heterostructure of just two materials) comprises a crystal of silicon-doped AlGaAs grown epitaxially on a crystal of GaAs (Figure 4.4(c)). Electrons are liberated from the silicon dopants in the AlGaAs and transfer into the GaAs. The ionised silicon dopants hold these free electrons up against the GaAs–AlGaAs interface and, as in the silicon MOSFET, a two-dimensional electron system is formed. These two-dimensional systems can be of extremely high quality owing to the crystalline purity and abruptness of the interface. Compound semiconductors such as GaAs are of technological importance since, unlike silicon, they can emit light efficiently. They are the basis for solid-state lasers and light-emitting diodes, which are essential for fast modern fibre-optic telecommunications, the Internet and compact disc players, *inter alia*. The semiconductor quantum cascade laser which operates at mid-infrared frequencies is one example of the versatility of modern compound semiconductor growth and its ability to engineer sophisticated optical devices. Compound semiconductor transistors also switch faster than do silicon MOSFETs and are exploited in mobile telephones.

Two pioneers of fast opto-electronic and microelectronic components based on layered semiconductors, Zhores Alferov and Herbert Kroemer, shared the Nobel Prize for Physics in 2000 (with Jack Kilby) for their crucial role in the foundation of modern information technology.

4.10 The fractional quantum Hall effect – an electron liquid

Two years after the discovery of the integer quantum Hall effect, Daniel Tsui and Horst Störmer of AT&T Bell Laboratories observed Hall plateaux and longitudinal resistance minima in low-temperature studies of the two-dimensional electron system formed in GaAs–AlGaAs heterojunctions grown by Arthur Gossard. They found not only Hall plateaux quantised at resistances of $25\,813/i\ \Omega$ with i an integer (the integer quantum Hall effect) but also two additional Hall plateaux quantised at $25\,813/i\ \Omega$ with i equal to $\frac{1}{3}$ and $\frac{2}{3}$. It was clear that, although these new features were phenomenologically similar to the integer quantum Hall effect, they had a different origin. The fundamental interactions between the electrons themselves were suggested to be responsible since these interactions were expected to be particularly strong in the high-purity GaAs–AlGaAs heterojunctions.

Theoretical understanding was provided by Robert Laughlin, who found that, when the bottom Landau level was a third or two thirds full, interactions between the electrons in the two-dimensional electron system caused them to correlate together and condense into a new state of matter – an electron liquid. Furthermore, in this condensed state, electrical conduction no longer took place by means of the usual electrons, but by means of new fractionally charged particles carrying one third of the normal electronic charge. This remarkable phenomenon was called the fractional quantum Hall effect and was a very exciting discovery since it represents a rare example of what is known as a macroscopic quantum state – a quantum mechanical entity that exists over a large distance, perhaps up to several millimetres.

Many more fractional quantum Hall effect states have subsequently been observed. Figure 4.3(c) shows the rich spectrum of integer and fractional quantum Hall effect structure observed in a subsequent investigation. The fractional quantum Hall effect is the signature of a completely unexpected macroscopic quantum phenomenon and earned Tsui, Störmer and Laughlin the 1998 Nobel Prize for Physics.

4.11 The electron crystal

Interest in the low-temperature properties of interacting electron systems extends back to a proposal by Eugene Wigner in 1934 for a dilute three-dimensional crystalline electron state. Although the electron concentration in GaAs–AlGaAs heterostructures is too high for such a 'Wigner solid' to form (the inherent quantum mechanical motion of the electrons will always be sufficient to shake the crystal apart irrespective of how much you cool the sample), a transition to an electron solid at high magnetic field is expected (Figure 4.6). The electrons are each confined to a progressively smaller area with increasing field and the system minimises its potential energy by forming a lattice structure called a magnetically induced Wigner solid. (A true two-dimensional Wigner solid can be observed without a magnetic field in the dilute electron system formed when electrons are suspended above the surface of a dielectric such as liquid helium.)

As we have seen, however, the application of a strong perpendicular magnetic field has other profound effects on two-dimensional electron systems; the integer quantum Hall effect and the liquid-like fractional quantum Hall effect can be formed and competition between the correlated states (the fractional quantum Hall effect versus the magnetically

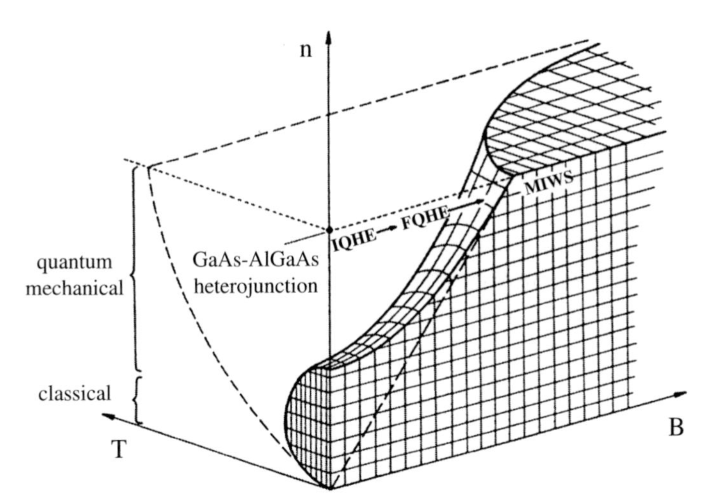

Figure 4.6. A schematic phase diagram for the electron solid (the hatched region). The axes are electron concentration (n), temperature (T) and magnetic field (B). At zero magnetic field (in the n–T plane), a classical Wigner solid is expected only at low temperatures for clean, dilute electron systems. If the electron concentration is too high (as is the case for the two-dimensional electron system found in GaAs–AlGaAs heterojunctions, for example), the electron's intrinsic quantum mechanical motion prevents the electrons from crystallising, even at the lowest possible temperatures. Under the influence of a magnetic field, the two-dimensional electron system passes first through the integer and fractional quantum-Hall-effect regimes (IQHE and FQHE), ultimately crystallising as a magnetically induced Wigner solid (MIWS).

induced Wigner solid) is of particular interest. Results of a variety of experiments have mapped the electron-liquid–solid phase boundary and have provided strong evidence that this solid phase indeed exists, but is broken into domains about one micrometre in size and pinned by residual impurities. Investigations of the electrical conduction of a pinned crystal can be problematic and so there has been interest in alternative optical measurement techniques such as magneto-photoluminescence to provide a local experimental probe of the electron crystal lattice structure.

4.12 Nanostructure Physics

Electrons can be constrained to patterned nanoscale geometries in the two-dimensional plane by etching vertically into the device or by imposing electrostatic confinement. Electrostatic patterning has the advantages of providing better resolution and a controllable degree of confinement. If the additional confinement is extreme (of the order of the electron wavelength), the energy of the two-dimensional electrons will become quantised in the confining direction and an electronic system of lower dimensionality such as a one-dimensional wire or a zero-dimensional box will form. The investigation of such systems has led to a wealth of new physics because the electrical, optical and thermal properties of electronic systems depend strongly upon their dimensionality. Electron-beam lithography is the key fabrication technique and uses the focused beam of a scanning electron microscope to write directly into an electron-sensitive resist. The exposed resist is chemically modified and can be selectively removed to allow deposition of a metal gate onto the semiconductor surface at submicrometre resolution. Under an appropriate bias, the two-dimensional electrons lying below and to the side of the gate structure are depleted and the remaining electrons are constrained to flow around the nanoscale geometry described originally by the beam. Electron-beam lithography is a very versatile technique; if you want to refine the pattern or even change the geometry completely, you simply modify the computer program that controls the beam.

The electrostatic lateral confinement of a two-dimensional system into a one-dimensional constriction was established in 1982 in *n*-type silicon MOSFETs by Michael Pepper and Alan Fowler, after Pepper had demonstrated the electrostatic squeezing of a three-dimensional electron system into a two-dimensional system in 1978. This technique was subsequently extended to produce very short constrictions in the GaAs–AlGaAs system and an exciting new phenomenon was discovered. A surface metal 'split gate' (Figure 4.7(a), upper inset) comprising two rectangular gates 1 μm wide and separated by 0.7 μm was fabricated by electron-beam lithography. A negative bias on these gates squeezed the two-dimensional electrons below into a variable-width one-dimensional wire. At low temperatures, the electrons could pass though the short one-dimensional constriction without scattering owing to the very high purity of the GaAs–AlGaAs structure. This 'ballistic' conduction is very different from customary diffusive electron transport, in which electrons repeatedly scatter off impurities and defects, and has some remarkable properties. Figure 4.7(a) shows the conductance of a recent ballistic one-dimensional wire. As the constriction becomes narrower, the conductance decreases.

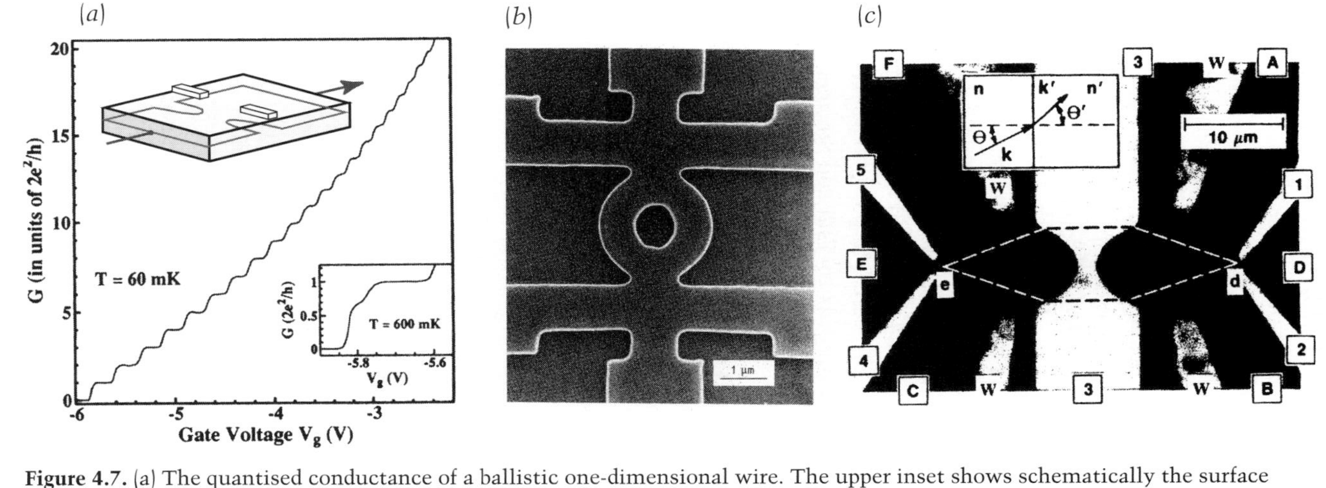

Figure 4.7. (a) The quantised conductance of a ballistic one-dimensional wire. The upper inset shows schematically the surface metal 'split gate' geometry comprising two rectangular gates of width and separation of the order of 1 µm. A negative voltage on these gates squeezes the underlying electrons into a variable-width one-dimensional wire in the gap between the fingers (the grey outline). As the constriction becomes narrower, the conductance G decreases, but it does so in quantized steps of $2e^2/h$. The lower inset shows the recently identified conductance feature at $0.7(2e^2/h)$, which is currently not understood. (From Thomas *et al.* 1996 *Phys. Rev. Lett.* 77, 135–8.) (b) A scanning electron micrograph image of an Aharonov–Böhm electron-interference ring. The relative phase of the electron waves travelling around the two arms of the ring can be tuned by application of a magnetic field. If the field is continuously swept, the electron waves periodically interfere constructively and destructively when they recombine, giving rise to oscillations in conductance. (From Ford *et al.* 1988 *J. Phys. C: Solid State Phys.* **21**, L325–31.) (c) A scanning electron micrograph image of the gate geometry of an electrostatic lens. Electrons fired from the left are diffracted in the region under the lens-shaped gate and focused into the constriction on the right. (From Spector *et al.* 1990 *Appl. Phys. Lett.* **56**, 1290–2.)

However, it does so in a step-wise manner with the conductance quantized at $2ie^2/h$, with i an integer. The reason for this is that, when the width of the constriction becomes comparable to the electron wavelength, there forms in the constriction a series of one-dimensional channels (similar to the edge-state channels in the quantum Hall effect), which transmit the electrons. It can be shown that each occupied one-dimensional channel contributes a fixed conductance $2e^2/h$, so that, as the width of the constriction changes and the number of occupied one-dimensional channels changes, the conductance steps up or down abruptly. (The factor of two difference between the conductance of the ballistic one-dimensional channel and the quantum Hall effect edge channel is caused by the presence of the magnetic field in the latter case.) A further plateau has been observed at $0.7(2e^2/h)$, which cannot be explained by conventional theoretical models (Figure 4.7(a), lower inset). Perhaps this feature will prove to be caused by interactions between the electrons analogous to Tsui and Störmer's first observation of fractional plateaux in studies of the integer quantum Hall effect.

The field of nanostructure physics has exploded over the last fifteen years with the investigation of a vast array of ingenious geometries – two are outlined here very briefly to give a flavour (see Beenakker and van Houten (1991) for further examples). Interelectron interference was investigated by constructing the ring structure in Figure 4.7(b). Quantum mechanical electron waves can interfere in the same way light waves interfere to give coloured patterns on oily puddles or soap bubbles. The relative phase of electrons travelling around the two arms of the ring depends upon the magnetic field passing through the middle (this is called the Aharonov–Böhm effect). By sweeping the magnetic field, the electron waves travelling along the two paths are repeatedly tuned in and out of phase so when they recombine they periodically interfere constructively and destructively, giving rise to oscillations in conductance of period $h/(eA)$ (A is the area of the ring). Figure 4.7(c) shows an electrostatic lens. The two-dimensional electron system is partly depleted under the lens-shaped gate, which has the effect of reducing the electron momentum in this region. Electrons fired from the left are diffracted and focused into the constriction on the right (note that this is achieved by having a structure shaped like an optically diverging lens).

4.13 The future

In the late 1960s, Gordon Moore proposed that the complexity of microprocessors (the number of components per unit area) would double every eighteen months. 'Moore's Law' has so far been obeyed. However, it is predicted that the prevailing silicon technology will not be susceptible to this exponential progression in miniaturisation (and associated circuit capability) for more than a further ten years. With the enormous capital investment and expertise already tied up in silicon technology, it is natural to develop existing proven technology. But even relatively direct developments (for example, changing circuit interconnects from aluminium to the better conducting copper, or pushing photolithography to progressively smaller wavelengths into the deep ultra-violet), require a large investment in time and money. Since a modern silicon fabrication plant costs around £1.5 billion (and this cost has increased exponentially with time – Moore's second law), there will come a point at which small improvements no longer outweigh the necessary investment. Furthermore, device dimensions are already approaching the limit at which quantum effects will interfere with their operation – the MOSFET oxide thickness has gradually been reduced to around 5 nm; at 2 nm, quantum mechanical tunnelling through the oxide impairs performance. Dissipation of heat from the ever-more densely packed transistors is becoming problematic too – reminiscent of one of the complaints about the vacuum tube before the transistor superseded it. Of course, this technology may simply saturate – the internal combustion engine has hardly developed in decades but people still buy cars. Other materials systems may prove to play a larger role such as GaAs (which, as we have seen, has some operational advantages over silicon, but the technology is expensive) or silicon–germanium heterostructures (which have the prospect of combining the flexibility of the layered GaAs-based devices with compatibility with silicon MOSFETs).

However, I want to look well into the future. Instead of trying to pattern slices of semiconductor at an ever-finer resolution, why not build upwards – assemble atoms and molecules individually into appropriate three-dimensional configurations? Furthermore, as devices become smaller and approach the quantum limit, quantum mechanical effects, rather than being deleterious, may become central to device operation. New electrical and optical characteristics will emerge and be exploited and the quantum mechanical phenomena prized by scientists will enter the public domain.

4.14 Molecular electronics

Researchers have been looking to fabricate circuit elements on the ultimate molecular scale for many years now. It is nearly thirty years since the first proposal was made for a molecule that would show rectification of electrical conduction though its molecular orbitals – a molecular crystal detector. A range of molecular attributes including electrical, optical, and mechanical properties, nuclear spin, conformation and lock-and-key recognition, *inter alia*, could be exploited for switching and memory functionality (for a review, see Petty *et al.* 1995). The potential of a molecular-based technology is that an astonishing quantity (on the scale of the Avogadro number, about 10^{23}) of identical components can be synthesised simultaneously. Since each will be a fraction of the size of even the smallest conventional semiconductor circuit element, extraordinary packing densities may be possible.

Progress in the study of individual molecules has already been made – for example, the conductance of a single benzene ring spanning the two gold faces of a break junction has been measured. Furthermore, molecules have been designed to mimic conventional electronic circuit elements and a molecular resonant tunnelling diode and a molecular rectifying diode have been demonstrated experimentally. Molecular circuits for the AND and OR digital logic gates have been proposed using appropriate arrangements of molecular rectifying diodes ('diode–diode' logic structures). By incorporating molecular resonant tunnelling diodes too, a molecular NOT digital logic gate can be constructed, thus creating the complete set of essential logic gates. By combining these gates one can, in principle, construct molecular structures to implement complex logic. However, there are a number of issues that must be addressed before such molecular circuits can be utilised. For example, when molecular-scale devices are combined together in a circuit they are unlikely to behave in the same way as they do in isolation. Their individual characteristics will be altered by the Coulomb and quantum mechanical interactions between one another.

Many molecules such as polyacetylene (the simplest conjugated polymer) become semiconductors upon polymerisation but can be 'doped' into becoming conductors using alkali metals or halogens to add electrons into, or remove electrons from, the π orbitals. Conducting and semiconducting polymers are already exploited for solar cells and large-area electroluminescent displays, for example, and have been demonstrated to be susceptible to the field effect in polymer-channel transistors. The discovery and development of conducting polymers was recognised by the Nobel Prize for Chemistry in 2000.

There is a growing interest in the electrical and mechanical properties of rigid-carbon structures such as C_{60} 'buckyballs' and, of particular relevance to molecular conductors, carbon nanotubes. These are cylindrical molecules (rather like seamless rolled-up sheets of graphite) which are about 1 nm in diameter and up to 100 μm long. Their electrical properties are sensitive to the tube diameter and to the helical pitch of the hexagonal carbon lattice along the tube, and they can be either metallic or semiconducting. Owing to their nanoscale diameter and the special electronic structure of graphite, single nanotubes have been shown to behave as one-dimensional quantum wires similar to the 'split-gate' structure discussed earlier. A field-effect transistor has been proposed comprising a single semiconducting nanotube spanning two metal electrodes with the substrate acting as the gate to modulate the current flow though the nanotube.

4.15 Biomolecular electronics and computers

Perhaps the most exciting prospect, though, involves one of the most fundamental molecules of life itself. The selective self-assembly and molecular recognition properties inherent to DNA (deoxyribonucleic acid) and embodied by Watson–Crick pairing might be exploited to engineer complex molecular networks with exotic electrical and optical properties. DNA has been used to organise colloidal particles into macroscopic crystal-like aggregates and to control the conformation of semiconductor nanoparticle assemblies – it is clear that this technique has the potential to self-assemble far more complicated structures. This property of DNA has already been exploited as the basis of a DNA 'computer', first demonstrated by Leonard Adleman to solve a version of the mathematical 'travelling salesman' problem: find the shortest route between seven cities without retracing your steps. A unique twenty-base strand of DNA was made to represent each of the seven cities and each possible road between them. The base sequences were arranged so that the 'sticky ends' of each road strand would connect to the two appropriate city strands according to the Watson–Crick binding rules of DNA. By mixing billions of copies of each strand together, all possible combinations (and hence routes between the cities) were generated. The unique solution to the problem was the one represented by the shortest DNA molecule containing each city sequence only once, and this was identified by carrying out a week-long series of biochemical reactions.

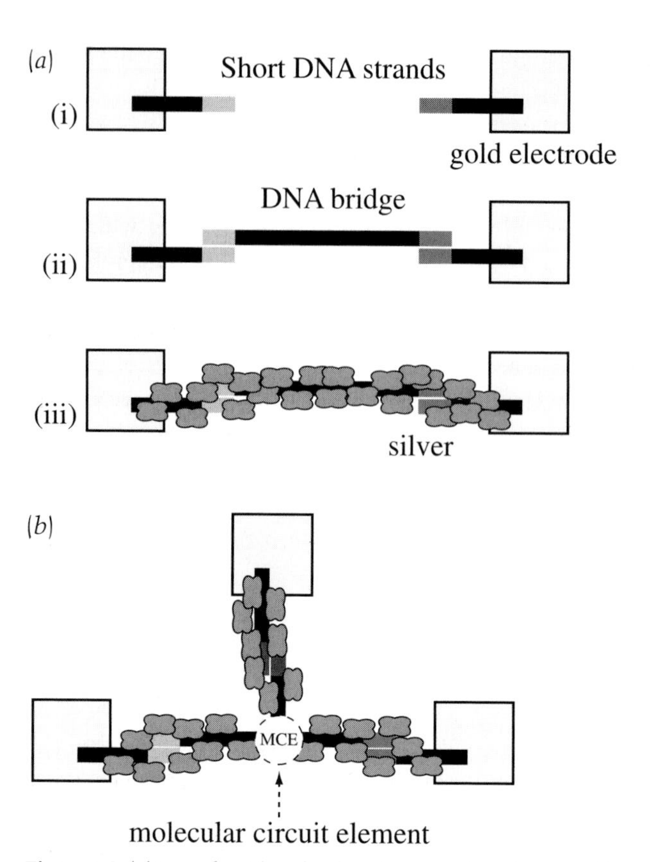

Figure 4.8. (a) Two short (twelve-base) pieces of single-strand DNA are attached to electrodes, each strand comprising a different base sequence (after Braun *et al.* 1998 *Nature* **391**, 775–8). These strands (and hence the electrodes) can be bridged by a third, longer, strand of DNA, each end of which is complementary to one of the sequences on the short strands. A highly selective localisation of silver ions along the DNA can be performed with the subsequent formation of metallic silver aggregates. DNA can thus be used to form a template that can nucleate a 100-nm-thick conducting silver wire. This is below the width achievable with standard, industrial processing technology. (b) This technique has the potential to bring about self-assembly of complex nanometre-scale conducting networks, incorporating functional molecular circuit elements.

The immense computational capacity of the DNA ligation reaction has been extended to solve other combinatorial problems by quickly producing billions of products representing all possible solutions. However, the selection of the correct solution is, at present, slow and laborious.

The self-assembly and molecular recognition properties of DNA may, however, address the nagging problem of how functional molecular circuit elements might be connected to each other and to the outside world. Recently, a series of experiments have been performed to measure the conduction properties of DNA itself and to investigate other ways of employing DNA to make electrical connections.

A recent experiment showed that a strand of DNA can form a template that can nucleate a 100-nm-thick conducting silver wire. Two gold electrodes, each about 1 μm square, were evaporated 12–16 μm apart on a glass slide and a short (twelve-base) piece of single-strand DNA was attached to each electrode. Each strand comprised a different base sequence (Figure 4.8(*a*)). These two strands of DNA, and hence the electrodes, were bridged by a third strand of DNA, 16 μm long, possessing two twelve-base sticky ends. Each end was complementary to one of the sequences on the short strands of DNA bound to the electrodes. A selective localisation of silver ions along the DNA was performed leading to the formation of silver complexes with the DNA bases. These complexes seeded metallic silver aggregates along the DNA skeleton to form, ultimately, a 100-nm-wide conducting granular silver wire connecting the two electrodes.

The conduction properties of the DNA molecule itself are currently the subject of debate. Not only are they of interest to physical scientists for the reasons discussed here, but biochemists and molecular biologists believe the ability of DNA to transport electrons may play a role in the *in vivo* repair of the DNA helix after radiation damage. Although indirect studies of DNA conduction have been performed by studying the electron-transfer-mediated quenching of fluorescent molecules attached to DNA strands, direct conductivity studies have recently been performed. One study investigated DNA 'ropes' at least 600 nm long, which comprised several associated DNA molecules, and found efficient conduction with a resistivity comparable to that of a conducting polymer or a good semiconductor. However, a second investigation studied the conduction of individual poly(G)-poly(C) DNA molecules which were 10.4 nm long (thirty base pairs) and found, by contrast, much poorer conduction.

Although there is clearly much work still to be performed, one day, entire three-dimensional microprocessors and memories might be created by using the self-assembly properties of DNA to build networks of molecular circuit elements (transistors, capacitors, diodes, etc.). These functional molecules would be interconnected electrically by DNA wires, DNA-assembled metallic wires or conducting conjugated molecules (Figure 4.8(*b*)).

4.16 Conclusions

I believe, however, that it is a mistake to focus exclusively on the way existing technology operates and to try to create alternative schemes that operate in essentially the same manner. New devices reveal new physics. New devices will have new methods of operation. Some of these are predicted and being sought; others will be discovered by serendipity. Shockley, Bardeen and Brattain tried to develop the field-effect transistor but instead discovered the bipolar transistor. Their work was grounded on years of fundamental experimental and theoretical research on crystal detectors, semiconductors and the 'esoteric' theories of quantum mechanics. If the concept of a self-assembled biomolecular computer seems far-fetched, remember that it can take a long time for ideas to become technologically feasible and reach fruition. Nearly a century passed between J. J. Thomson's original search for the field effect and it finally achieving widespread use in the microprocessor – but it happened.

4.17 Further reading

Beenakker, C. W. J. and van Houten, H. 1991 *Solid State Physics Volume 44: Advances in Research and Applications* edited by Ehrenreich, H. and Turnbull, D. New York: Academic Press.

Davies, A. G. 2000 Quantum electronics: the physics and technology of low-dimensional electronic systems into the next millennium, *Phil. Trans. Roy. Soc.* A **358**, 151–72.

Petty, M. C., Bryce, M. R. and Bloor, D. 1995 *Introduction to Molecular Electronics*. London: Edward Arnold.

Riordan, M. and Hoddeson, L. 1997 *Crystal Fire – The Birth of the Information Age*. New York: Norton.

Seitz, F. and Einspruch, N. G. 1998 *Electronic Genie – The Tangled History of Silicon*. Illinois: University of Illinois Press.

Shurkin, J. 1996 *Engines of the Mind*. New York: Norton.

5 Spin electronics

Michael Ziese

Department of Superconductivity and Magnetism, University of Leipzig, Linnéstraße 5, 04103 Leipzig, Germany

5.1 Introduction

Spin electronics is a new field of research in condensed matter physics and both experimental and theoretical understanding is in the initial stages. The development of spin electronics was stimulated by the current trend towards miniaturisation in the electronics industry that will reach a limit in the near future. This might lead to the replacement of semiconductor devices by magnetic components in future information technology hardware. At typical doping levels, the number of charge carriers in semiconductor nanostructures becomes small, eventually approaching the limit of single electron devices. Since carrier densities in metals are much larger than those in semiconductors, metallic devices based on macroscopic principles of operation can be more favourably miniaturised. In semiconductors currents are carried by electrons or holes (depending on doping – substitution by various chemical elements) and the basic operational feature of semiconductor devices is the differential manipulation of these two types of currents. In direct analogy it is possible to label electrons by their spin orientation and thus spin electronics uses the differential manipulation of currents with defined spin polarisations as the fundamental principle for operation of devices. This principle is sensible only if the spin information can be transported over reasonable length scales in solid-state devices. It was experimentally found that this is, indeed, possible and that the spin-diffusion length – the length an electron can travel without losing its spin memory – is several micrometres in some metals. Since new electronic devices will have to be fabricated on nanometre scales in order to compete with the existing semiconductor technology, the spin-diffusion length is very long indeed compared with a single electronic element.

At present, there are various approaches to spin electronics, depending on the scientific background of the individual research group: investigation of (a) metallic systems, (b) oxide nanostructures and (c) magnetic semiconductor devices. This article presents an overview of spin electronics based on oxides and gives only brief comments on other developments.

The sources of spin-polarised electrons are itinerant ferromagnets. The energy bands in ferromagnets depend on the electron spin such that the energy levels for the two spin directions are shifted with respect to each other. Since the energy levels are filled hierarchically, in a ferromagnet there are more electrons of one spin direction present. The electrons are called majority and minority carriers, respectively. The magnetisation of the ferromagnet is due to this spin imbalance and currents drawn from such a material are also spin-polarised.

An early success story of spin electronics is provided by the discovery of giant magnetoresistance in metallic multilayers in the late 1980s that led to the development and mass production of read-heads for hard disks based on this effect in less than ten years. These read-heads consist of multilayers composed of ferromagnetic layers separated by noble-metal interlayers with the electrical current flowing parallel to the layers. The roadmap for industrial development of hard disk drives envisages new generations of read-heads based either on spin valves or on tunnel junctions. These are basically composed of two ferromagnetic layers separated by a noble metal or insulating layer, respectively; the electrical current flows perpendicular to the layers. Spin valves and tunnel junctions are also expected to form the building blocks of a two-terminal, non-volatile magnetic random access memory (MRAM). For further reading see de Boeck and Borghs (1999). Advantages of elemental ferromagnets are that they have high Curie temperatures and advanced techniques for preparation of thin films are available; the main disadvantage is the moderate degrees of spin polarisation of elemental ferromagnets such as iron (44 per cent), nickel (11 per cent) and cobalt (34 per cent).

An important development in the field of spin-polarised transport came with the rediscovery of colossal magnetoresistance in the manganites of the type $La_{0.7}Sr_{0.3}MnO_3$. The magnetic and electrical properties of these compounds were first investigated in the early 1950s. However, the fabrication of technologically important thin films and the observation of very large magnetoresistance values was first reported at the beginning of the 1990s. The magnetoresistance in the manganites was labelled 'colossal magnetoresistance' in order to distinguish it from the giant magnetoresistance observed in metallic multilayers. The manganites have been studied intensively in recent years and rapid progress has been made, both in the fabrication of high-quality films and in gaining theoretical understanding of the transport mechanism. It was, however, soon realised that the intrinsic colossal magnetoresistance is not technologically relevant, since large changes in resistance are obtained only in high magnetic fields, of the order of several teslas. Therefore, recent research activity has been focused on extrinsic magnetoresistance effects, namely effects related to electron tunnelling and spin dependent scattering near grain boundaries, in the hope of engineering structures with a high sensitivity to magnetic fields. The extrinsic magnetoresistance is intimately linked to the spin-polarisation and this was found to be very large, approaching 100 per cent, thus giving the manganites a much greater potential for use in spin-electronic devices than that of conventional ferromagnets. The main disadvantage of the manganites, for room-temperature applications, is the relatively low Curie temperature, at most 380 K. Alternative materials are the oxide ferromagnet CrO_2 with a Curie temperature of 400 K and the ferrimagnet Fe_3O_4, with a relatively high Curie temperature of 858 K. Both these compounds are believed to be 100 per cent spin-polarised, although they exhibit only a small intrinsic magnetoresistance.

Oxide films are grown epitaxially by various deposition methods such as pulsed laser deposition, magnetron sputtering and molecular beam epitaxy on single-crystal substrates such as $SrTiO_3$ and $LaAlO_3$. The preparation techniques for manganite films are quite advanced and epitaxial manganite films have single-crystal quality; Fe_3O_4 and CrO_2 films are difficult to fabricate with reasonable quality. Manganite and magnetite films have been patterned using optical lithography on a micrometre scale; patterning on a submicrometre scale has not yet been reported and the achievement of high quality nanostructures will present a formidable technological challenge. One advantage of manganite film technology is the possibility of integrating high-temperature superconductors that have perovskite structures with similar lattice constants. The first experiments on such devices have allowed the study of the superconducting state by the injection of spin-polarised currents from an adjacent magnetic layer. In this article the path of spin electronics from basic physics to device applications and a future generation of information technology devices is followed.

5.2. Colossal magnetoresistance in magnetic perovskites

The interest in magnetic perovskite materials of the type $La_{0.7}AE_{0.3}MnO_3$ was revived after the discovery of a large magnetoresistance in thin films of ferromagnetic $La_{0.7}Ba_{0.3}MnO_3$ in the early 1990s. AE stands for an alkaline earth element such as Ca, Sr or Ba. The materials were originally synthesised and studied in polycrystalline form in the 1950s at the Philips Research Laboratories. In this section, recent experimental and theoretical studies of the transport properties of the manganites are briefly reviewed. For an extensive review see Coey *et al.* (1999).

In Figure 5.1(a) the resistivity (left-hand axis) in zero field and an applied magnetic field of 1 tesla, the magnetisation (right-hand axis) and in Figure 5.1 (b) the magnetoresistance ratio, defined as the deviation from the zero field resistivity ρ_0 normalised with respect to this value,

$$\mathrm{MR} = [\rho(H) - \rho_0]/\rho_0,$$

of an epitaxial $La_{0.7}Ca_{0.3}MnO_3$ film of thickness 170 nm are shown as functions of temperature. The film is clearly ferromagnetic with a Curie temperature of 225 K. The resistivity exhibits a maximum near the Curie temperature; below this temperature the resistivity is metallic, above it the film is semiconducting: this means that the ferromagnetic transition is accompanied by a metal–insulator transition. This is a consequence of the peculiar double-exchange mechanism that is characteristic of the manganites. The magnetoresistance is sharply peaked at the Curie temperature and has a maximum value of 42 per cent in a magnetic field of 1 T. The field sensitivity, however, is small, with a maximum value of 0.2 per cent mT^{-1}. For comparison, in metallic multilayers typical values exceed 1 per cent mT^{-1} at room temperature. Since most applications require a large sensitivity in magnetic fields of a few milliteslas, the intrinsic magnetoresistance of manganites is hardly competitive to that of giant-magnetoresistance structures.

The metallic character of the manganites arises from electron transfer between manganese ions via oxygen ions. According to the double-exchange model this is possible only if the spins of the individual manganese ions are parallel. Thus metallic conductivity and ferromagnetism are

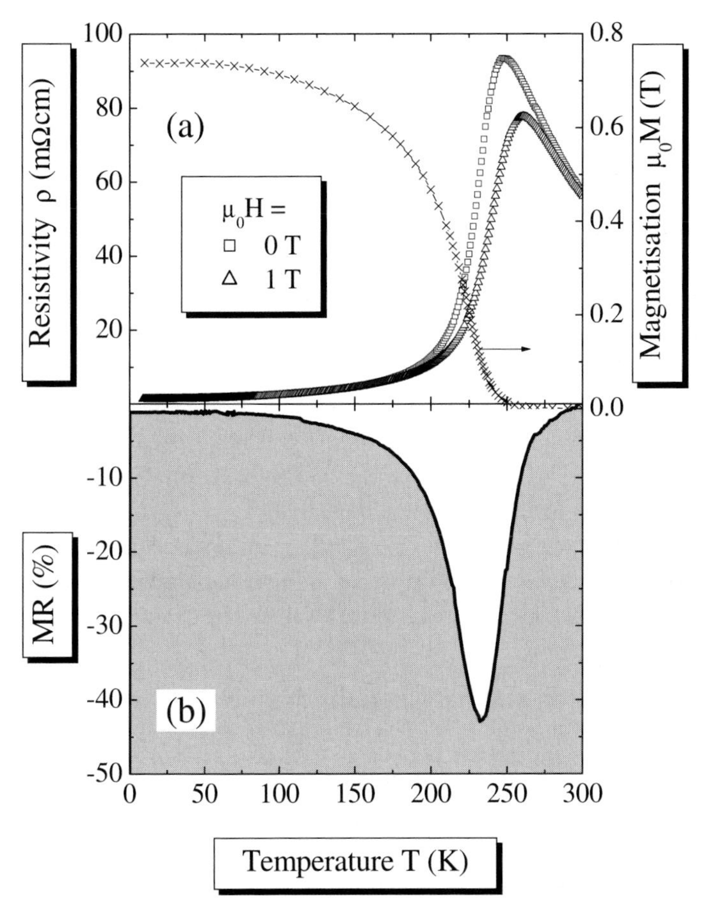

Figure 5.1. This figure shows some basic experimental data regarding a single-crystal film of $La_{0.7}Ca_{0.3}MnO_3$ on $LaAlO_3$. In (a) the electrical resistivity ρ (left-hand axis) measured in zero field (squares) and an applied magnetic field of 1 T (triangles) is compared with the magnetisation M (right-hand axis, crosses). The magnetisation exhibits a spontaneous moment below the Curie temperature of 225 K, indicating the ferromagnetic character of the compound. At this temperature the resistivity undergoes a transition from a high-temperature semiconducting to a low-temperature metallic phase. The application of a magnetic field suppressed the semiconducting phase and led to a considerable negative magnetoresistance MR as shown in (b). Hence the name 'colossal-magnetoresistance' manganites for these manganese-based oxides. MR is defined as the normalised deviation of the resistivity from its zero-field value.

intimately linked. Application of a magnetic field leads to a greater alignment of the manganese spins and thus to an increase in conductivity, especially in the vicinity of the Curie temperature; this is in qualitative agreement with experiment.

Recent theoretical investigations showed that the double-exchange model proposed in the 1950s is not sufficient to explain the colossal magnetoresistance. These studies indicate that the carriers interact strongly with lattice vibrations (phonons). The strong electron–phonon coupling leads to the formation of polarons above the Curie temperature; these are electrons accompanied by a large lattice distortion. The polarons certainly have magnetic character, leading to self-trapping, and the metal–insulator transition could be induced by the unbinding of trapped polarons. Although the existence of polarons above the Curie temperature is well supported by transport measurements, the nature of the metal–insulator transition and the mechanism for colossal magnetoresistance have not yet been clarified fully. The manganites are an important system for the study of transport properties under the influence of a strong electron–phonon interaction, since the coupling strength can be varied by chemical doping.

5.3. Spin polarisation and half-metallic magnets

The concept of spin polarisation plays a central role in spin electronics. Intuitively, it means an imbalance between majority carriers with their spin parallel to the magnetisation and minority carriers with their spin anti-parallel. It recently became clear, however, that spin polarisation is not a universal characteristic property of a ferromagnet, but rather depends sensitively on the relation of the material to its surroundings, especially on the type of measurement used to determine its value. This creates the chance to engineer superstructures with well-defined individual spin polarisations. In strong ferromagnets such as nickel, the electronic *d* bands are known to be spin polarised such that the majority band lies below the Fermi level, whereas the minority band crosses the Fermi level. However, the *s* bands are not spin-split, leading to a finite density of states at the Fermi level in both sub-bands. This is schematically indicated in Figure 5.2(a) showing the density of states $D(E)$ as a function of the electron energy E for majority and minority carriers, respectively. The electron energy levels are occupied up to the Fermi energy E_F and the transport properties of the material are determined by the electrons in the immediate vicinity

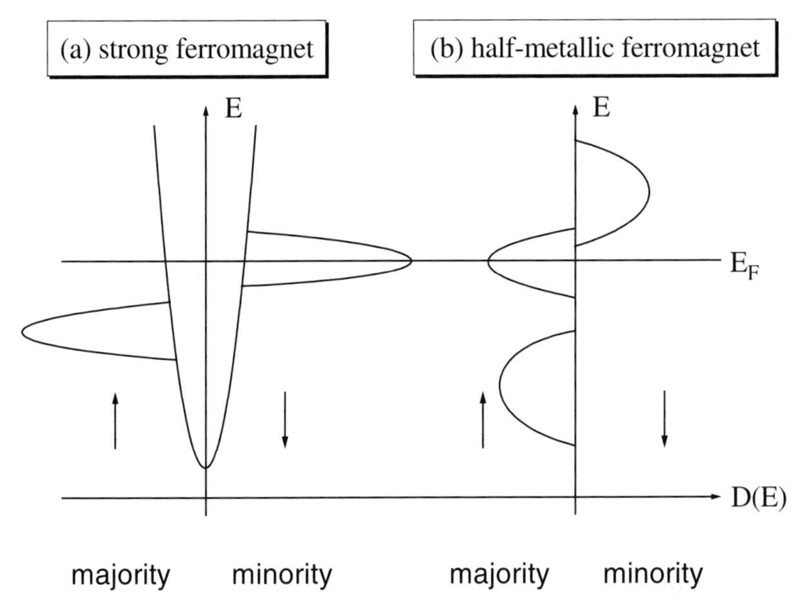

Figure 5.2. The density of states $D(E)$ as a function of the electron energy E for (a) a strong ferromagnet and (b) a half-metallic ferromagnet. The available single-electron energy levels are occupied up to the Fermi energy E_F. The broad density-of-states curve in (a) is derived from the extended *s* states, whereas the large narrow density-of-states peaks arise from the *d* bands. The ferromagnetic instability leads to a relative energy shift of the *d* levels in order to minimise the total energy and thus to the formation of a spontaneous magnetisation. The density-of-states curves in (b) are derived from *d* states. The striking thing about this diagram is the absence of minority electron states at the Fermi level, leading to a spin polarisation of 100 per cent.

of this energy. The simplest way to define the spin polarisation, P, is as the normalised difference between the densities $D_\uparrow$ and $D_\downarrow$ of the majority ($\uparrow$) and minority ($\downarrow$) electrons at the Fermi energy,

$$P=(D_\uparrow - D_\downarrow)/(D_\uparrow + D_\downarrow).$$

This is a definition that can be probed by a spin-polarised spectroscopy experiment measuring differences in the densities of states. In a transport measurement, however, a different kind of spin polarisation is measured. Since transport measurements always involve a flow of current, the spin polarisation of the electrical currents is relevant in this case and the corresponding expressions involve both the densities of states and the electron velocities. For further reading see Mazin (1999).

Another complication arises if the spin polarisation is measured using some kind of heterostructure such as a ferromagnet–insulator–ferromagnet or a ferromagnet–insulator–superconductor tunnelling junction. In this case the spin polarisation is found to depend on the barrier material. Consider cobalt as an example of a strong ferromagnet with a density of states as shown in Figure 5.2(a). From Figure 5.2(a) and the definition of the spin polarisation given above, a negative value of the spin polarisation is expected. The value measured in a $Co/Al_2O_3/Al$ junction, however, is *positive*, with a value of +34 per cent. If $SrTiO_3$ is used as the barrier material, a negative spin polarisation in agreement with the band structure is found. This has been a long-standing puzzle and is only being resolved at the moment. Numerical simulations indicate that the spin polarisation depends on the character of the interfacial bonds, with bonds between s orbitals favouring positive spin polarisation and d–d bonds favouring negative spin polarisation. A full understanding has not been achieved yet, but it is clear that this behaviour presents scientists with an additional degree of freedom for the design of their devices.

The new concept of half-metallicity was introduced on the basis of band-structure calculations. The striking feature emerging from those calculations was metallic behaviour for the majority carriers, coexisting with semiconducting, or insulating, behaviour of the minority carriers; this was indicated by a gap in the band structure for the minority carrier. A schematic diagram of the density of states of a half-metallic ferromagnet is shown in Figure 5.2(b). Half-metallic ferromagnets have a spin polarisation of 100 per cent within any sensible definition. Since only d states are near the Fermi level, it is hard to imagine that a change in sign of the spin polarisation can be achieved for some barrier–ferromagnet combination as was the case for elemental ferromagnets. The high spin polarisation makes half-metallic magnets the ideal candidates for use as sources or filters for spin-polarised currents in spin-electronic devices. The spin-diffusion length in half-metallic ferromagnets is certainly very large, since elastic spin-flipping from majority into minority states cannot occur. Band-structure calculations showed CrO_2, Fe_3O_4 and the manganites to be half-metallic. The experimental verification of this property, however, has proved to be difficult. Experiments based on Andreev reflection in superconductor–ferromagnet junctions yielded spin polarisation values of +85 per cent and +90 per cent at 4.2 K for $La_{0.7}Sr_{0.3}MnO_3$ and CrO_2, respectively. Measurements of the magnetoresistance of ferromagnet–insulator–ferromagnet junctions indicated a spin polarisation of about +83 per cent for $La_{0.7}Sr_{0.3}MnO_3$ at 4.2 K. Recent Fermi-surface measurements confirmed band-structure calculations and indicated that there is a high spin polarisation in the hole-like Fermi-surface sheets derived from the d bands.

5.4. Spin-dependent devices

5.4.1 Magnetic sensors based on grain-boundary magnetoresistance

Much research activity has been devoted to the study of extrinsic magnetoresistive effects arising near grain boundaries in the manganites. These effects are large, since the double-exchange mechanism is very sensitive to the local crystallographic order. It was hoped that the low-field response near grain boundaries could be tailored to yield devices with a reasonable sensitivity. Several device geometries, e.g. single grain boundaries grown on bi-crystalline substrates and step–edge junctions, have been investigated. Generally it is found that the resistivity of polycrystalline samples is strongly enhanced and exhibits a broad maximum below the Curie temperature, whereas the magnetisation exhibits essentially the same temperature dependence as that of epitaxial films. The magnetoresistance increases with decreasing temperature. The low-field magnetoresistance of a step–edge array at various temperatures is shown in Figure 5.3. The step–edge array was fabricated by growing a 25-nm-thick film of $La_{0.7}Ca_{0.3}MnO_3$ on a $LaAlO_3$ substrate that was etched into a step–edge pattern. This consisted of 200 steps along [110] that were 140 nm high and 20 μm apart. Grain boundaries grow near the edges. In small magnetic fields the magnetoresistance ratio is, indeed, much larger than that of epitaxial films, reaching a maximal magnetic field sensitivity of about 0.8 per cent mT^{-1}.

The grain-boundary transport mechanism is related to spin-polarised tunnelling; in this case, however, the barrier contains a considerable amount of impurity states, leading to inelastic tunnelling. Although a deep insight into the transport properties of the manganites was gained from the investigation of grain-boundary transport, it is very unlikely that grain-boundary devices will ever be applied as magnetic field sensors, since competitive magnetic field sensitivities are achievable only below 200 K.

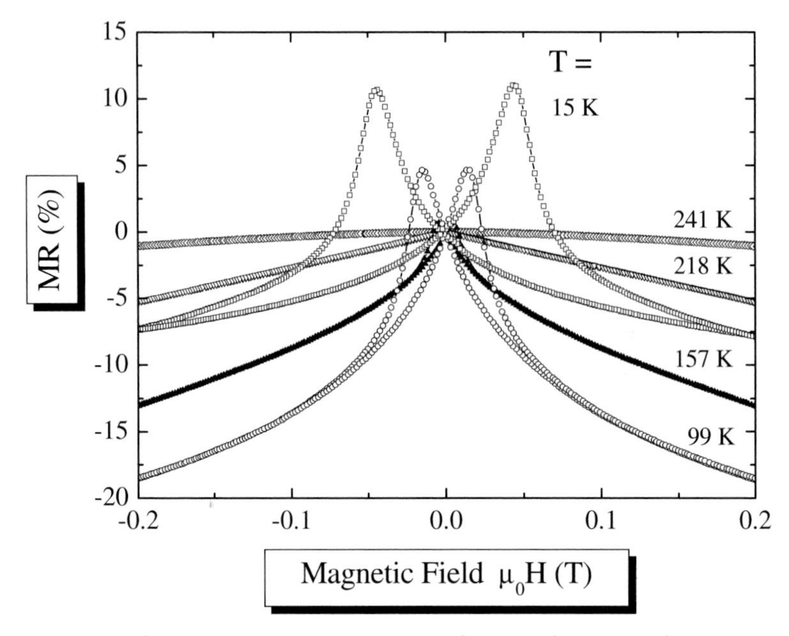

Figure 5.3. The magnetoresistance ratio MR of a step–edge array in low magnetic fields at various temperatures. The current flows across the steps. The array consists of 200 steps along [110], which are 140 nm high and 20 μm apart. In comparison with the epitaxial film shown in Figure 5.1, this step–edge array exhibits a significantly enhanced low-field magnetoresistance.

5.4.2. Ferromagnetic tunnelling junctions

Ferromagnet–insulator–ferromagnet tunnelling junctions have great potential for use as very sensitive magnetic-field sensors. A schematic diagram of such a junction is shown in Figure 5.4. To a certain approximation, the tunnelling magnetoresistance is determined only by the spin polarisation. Electrodes fabricated from half-metallic ferromagnets are particularly interesting, since the simple theory then predicts a diverging magnetoresistance. Real systems seldom exhibit infinite values, but the magnetoresistive effect can undoubtedly become very large. However, the techniques for fabrication of devices from oxide materials are not yet sufficiently advanced to produce high-quality tunnelling junctions. Figures 5.5(b) and (c) show the magnetoresistance at 4.2 K of two $La_{0.7}Sr_{0.3}MnO_3/SrTiO_3/La_{0.7}Sr_{0.3}MnO_3$ tunnel junctions fabricated by the IBM group. Whereas the larger junction displays only a moderate magnetoresistance of about 33 per cent, the small

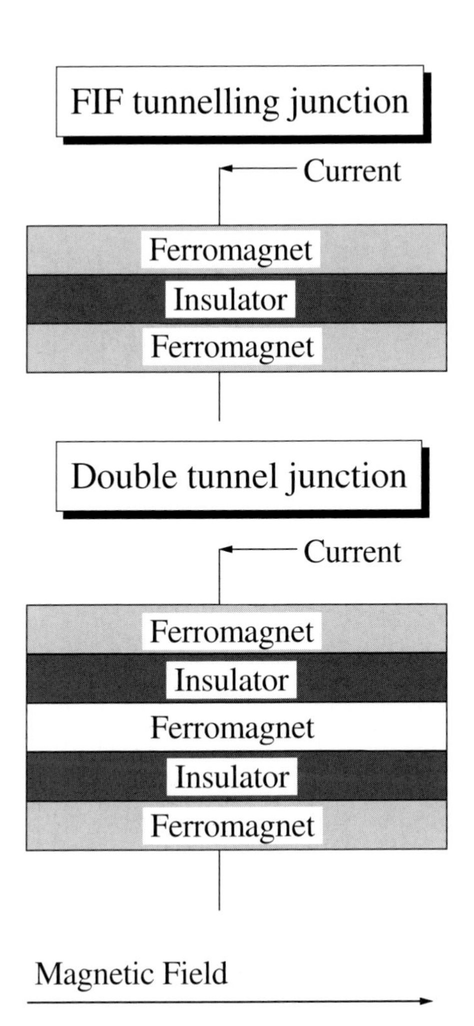

Figure 5.4. Schematic diagrams of a ferromagnet–insulator–ferromagnet (FIF) tunnelling junction and a double-tunnelling junction. The electrical current flows perpendicularly to the layers. These junctions are usually prepared from multilayer films using standard photolithography, etching and lift-off techniques.

junction exhibits a large magnetoresistance response of about 200 per cent. In Figure 5.5(a) the magnetic hysteresis loop of a single $La_{0.7}Sr_{0.3}MnO_3$ film is shown. The magnetoresistance maxima occur at the coercive field; different coercive fields of the two ferromagnetic electrodes in the trilayer structure are induced by variation in thickness. The magnetoresistance decreases rapidly with increasing temperature due to magnon scattering, inelastic tunnelling processes in the barrier, pinholes and a decrease in interfacial spin polarisation.

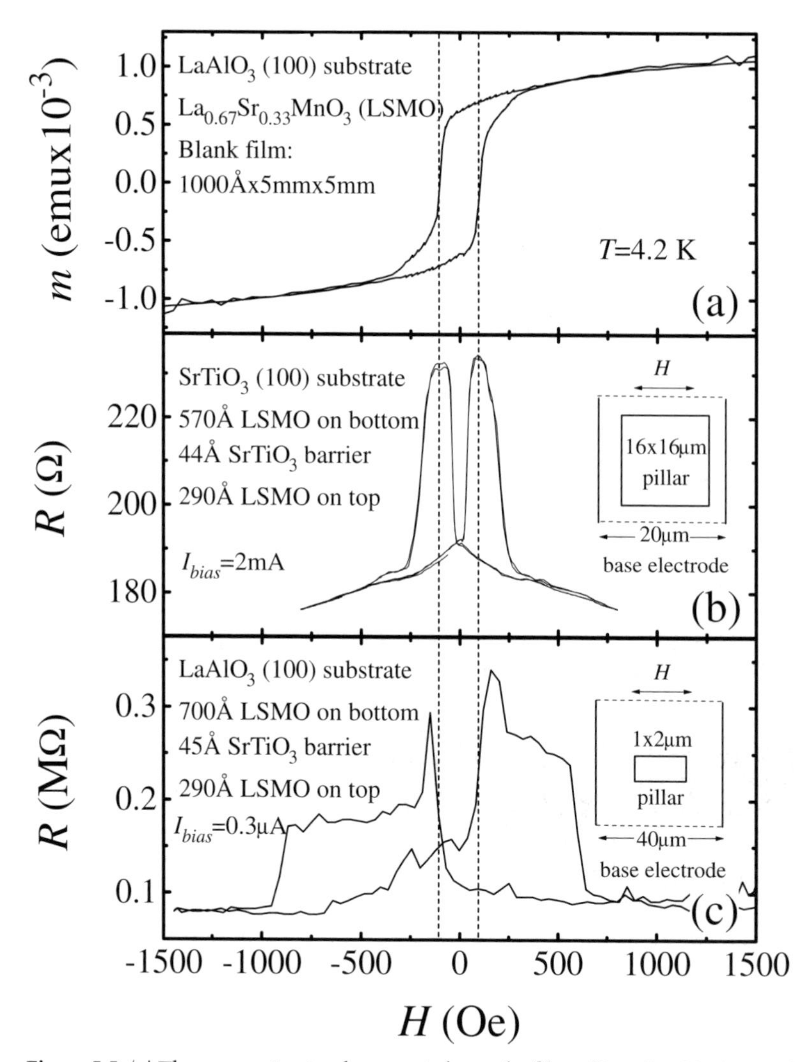

Figure 5.5. (a) The magnetisation hysteresis loop of a film of $La_{0.7}Sr_{0.3}MnO_3$ at 4.2 K. The coercive fields are defined by the zero crossings of the magnetisation. The resistances of (b) a large-area and (c) a small-area $La_{0.7}Sr_{0.3}MnO_3/SrTiO_3/La_{0.7}Sr_{0.3}MnO_3$ tunnelling junction are shown as functions of the magnetic field at 4.2 K. Whereas the large-area tunnelling junction exhibits a magnetoresistance of about 33 per cent, the resistance of the small-area junction changes by a dramatic 200 per cent during the magnetic-field sweep. (Reproduced from Sun (1998).)

Initial experiments on double-tunnelling junctions, shown schematically in Figure 5.4(b), have recently been performed. In such junctions the charging of the intermediate ferromagnetic island in the Coulomb-blockade regime is suppressed by the large Coulomb energy; electron transport occurs via simultaneous tunnelling of electrons to and from the intermediate electrode. This leads to a magnetoresistance with an even stronger divergence, being 'quadratically infinite', than in the case of single tunnelling junctions with half-metallic electrodes.

5.4.3 Injection of spin-polarised current into high-temperature superconductors

High-temperature superconductors such as $YBa_2Cu_3O_7$ react very sensitively to the injection of a spin-polarised current from a manganite layer. In a typical experiment the critical current of a superconducting film is found to decay strongly as a function of the injected spin-polarised current density. This result is tentatively attributed to pair breaking brought about by a large density of non-equilibrium quasiparticles.

5.4.4. Spin transistors

A spin transistor is a three-terminal device employing the differential manipulation of majority and minority carriers. A simple trilayer device using the conventional ferromagnets Co and permalloy ($Ni_{80}Fe_{20}$) was realised by Johnson (1993). Spin transistors based on oxide materials have not yet been fabricated, so in this section a diversion into the field of metallic multilayers and conventional spin electronics is made.

The geometry of the spin transistor is shown in Figure 5.6. It consists of a normal metal film, N, sandwiched between two ferromagnetic films, F1 and F2. A current is driven from the ferromagnetic electrode, F1, and is drained through the base, N. This current induces an accumulation of spin, i.e. a non-equilibrium magnetisation in the base. The density of majority carriers in the base is enhanced, whereas the density of minority carriers is reduced; this is equivalent to a shift in the chemical potentials of the two types of carrier. The collector electrode, F2, probes the chemical potential with respect to a reference voltage given by a non-magnetic metal probe. If the magnetisation in the electrodes F1 and F2 is parallel, the Fermi level in F2 aligns with the chemical potential of majority carriers in the base, whereas it is aligned with the minority Fermi level in the anti-parallel state. The collector voltage, being the difference between the chemical potentials for majority and minority spins in the base, switches between

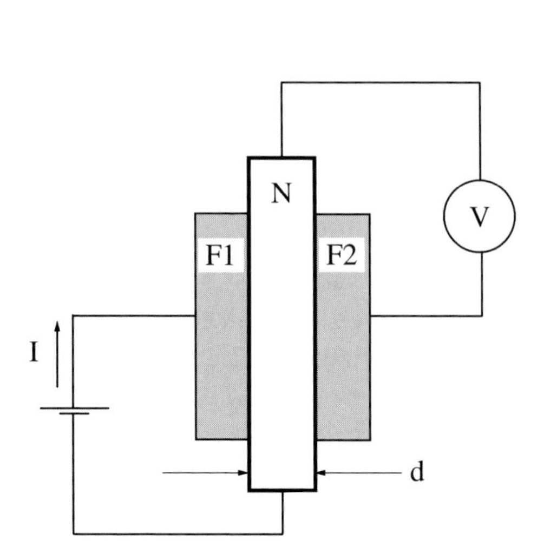

Figure 5.6. This figure shows the geometry of the spin transistor. A normal metal electrode, N, is sandwiched between two ferromagnetic electrodes, F1 and F2, made from ferromagnetic compounds with different coercive fields. During a magnetic-field sweep the output voltage of this device switches from a positive value when the magnetisations of the electrodes are parallel to a negative value when they are anti-parallel. A bi-stable output in zero field can be achieved, if the field sweep is halted before the larger coercive field is reached.

positive and negative values, thus realising a device with a bi-stable output (a memory cell).

The device operates as follows. Two ferromagnetic electrodes with different coercive fields are used such that, for an applied magnetic field in the range between the coercive fields, the magnetisation of the electrodes is anti-parallel. As a function of the applied field, the output voltage switches from a positive to a negative value and back while one is sweeping the field. A bi-stable voltage output can be realised in zero field, if the magnetic-field sweep is halted before the larger coercive field is reached and the magnetic field is reduced to zero.

The amazing feature of this spin transistor is the fact that a simple analysis indicates that the output signal will grow without bound on decreasing the volume of the base electrode. This makes these spin transistors competitive with semiconductor devices in the course of further miniaturisation. A rigorous mathematical treatment, however, showed that there is an upper bound to the output signal that is determined by the properties of the ferromagnetic electrodes as well as the transparency of the interface. The great advantage of half-metallic ferromagnets is the circumstance that this upper bound is diverging, thus offering the potential of large output signals from very small spin transistors. Experiments on this device using half-metallic ferromagnets have so far not been reported. This is probably due to the fact that the metallic base electrode has to be made from an oxide in order to preserve epitaxy throughout the structure. Since oxides tend to be 'bad metals', the spin-diffusion length in such a material might be quite short, imposing severe requirements on the microfabrication techniques. Here some interesting experiments are to be performed in the future.

5.5. Conclusion and outlook

In recent years astonishingly rapid progress has been made both in the fundamental study and in the application of spin-polarised transport. Many concepts have emerged and are actively being investigated at the present time. Studies of conventional metallic systems led to the discovery of giant magnetoresistance in multilayers and to the development of high-quality spin-polarised tunnelling devices using conventional ferromagnets. The devices are already being used or currently being engineered for use as magnetoresistive read-heads for hard disks and are obviously highly relevant commercially.

Future developments are likely to occur in the following fields.

5.5.1 Fundamental physics

This review shows that extensive progress has been made, covering a wealth of phenomena from fundamental transport processes in strongly correlated systems to single-electron-tunnelling processes in double junctions. These studies will continue in the near future. The magnetic oxides are complex compounds requiring an elaborate fabrication process similar to that for high-temperature superconductors. The history of research into high-temperature superconductivity shows that the investigation of transport properties is intimately linked to the quality of crystals, and that each generation of improvement in quality of materials reveals new fundamental physical properties. The same process is likely to occur during the course of studies on colossal magnetoresistance.

Investigation of spin-polarised electron tunnelling in oxide heterostructures is in its initial stages. Many problems have been identified and the fabrication of tunnelling devices will be improved by advances in the fabrication techniques as well as the introduction of new concepts. The theoretical understanding of spin-polarised tunnelling – and of giant magnetoresistance – is far from complete and the influence of the band structure on the process of tunnelling has to be clarified. Furthermore, the prospect of the manipulation of single electrons with well-defined spin orientation in double tunnel junctions opens up an interesting area of research.

5.5.2 Electronic devices

Obvious applications of ferromagnetic-oxide tunnel junctions are their operation as magnetoresistive read-heads for the next generation of storage media and their use in MRAM devices. These aims will certainly be sought; advances in multilayer fabrication techniques as well as lithography on the nanometre scale will be required. The fundamental problem with many oxide ferromagnets is their relatively low Curie temperatures; intense research in order to find magnetic oxides with higher Curie temperatures is under way.

The further development of spin electronics requires an active element with power gain comparable to that of a conventional semiconductor transistor. Since the amplification in semiconductor transistors is based on the nonlinearity of the current–voltage characteristics, it might be worthwhile to search for new spin transistor designs along these lines. The investigation of double-tunnel junctions with suitably tailored barrier materials seems to be most promising. A further requirement is the integration of spin-electronic devices into conventional semiconductor technology. One step in this direction has been made with the successful fabrication of high-quality manganite films on buffered silicon substrates. It would, however, be far more interesting to achieve the injection of spin-polarised carriers into semiconductors and create new devices based on spin-polarised electron and hole currents.

Recent research activity is actually starting to focus on spin-polarised transport in semiconductor devices. Here injection of spin into the semiconductor presents a great challenge. This problem is approached from two directions, namely (1) investigating injection of spin from Schottky contacts made of ferromagnetic metals and (2) facilitating injection of spin from a magnetic semiconductor epitaxially grown on the semiconductor being studied. Whereas the degree of injection of spin in the first class of structures is small, in the second class spin-injection efficiencies of up to 90 per cent at low temperatures have been reported. With appropriate fine tuning of the characteristics of magnetic semiconductors it should be possible to achieve injection of spin at room temperature. This would make available the possibility of an immediate integration of magnetic devices into existing semicondutor technology.

The development of working spin-electronic devices is expected in the near future. These might, however, not find an everyday use, but are more likely to be restricted to specific, high-performance applications such as supercomputing. One might think of the integration of MRAM devices based on magnetic oxides with rapid single flux-quantum logics based on high-temperature superconductors.

5.5.3 Research instrumentation

Scanning tunnelling microscopy relies on tunnelling of electrons between an atomically sharp tip and a clean surface. It facilitates the imaging of topographic surface features with atomic resolution. A scanning tunnelling microscope combined with spin-polarised tunnelling can be developed into a powerful tool for the imaging of magnetic structures with a resolution down to the atomic scale. Such a spin-polarised scanning tunnelling microscope is required for the investigation of magnetic domains in structures such as spin-electronic devices and magnetic-storage media on the nanometre scale. Furthermore, it could be used to measure the magnetic moments of single atoms, a feature yielding high potential for fundamental research in surface magnetism as well as for application in the technology of storage media.

5.6 Further reading

Coey, J. M. D., Viret M. and von Molnár, S. 1999 Mixed-valence manganites. *Advances in Physics* **48**, 167–293.

de Boeck, J. and Borghs, G. 1999 Magnetoelectronics. *Physics World* **12**, 27–32.

Fontcuberta, J. 1999 Colossal magnetoresistance. *Physics World* **12**, 33–6.

Johnson, M. 1993 Bipolar spin switch *Science* **260**, 320–2.

Mazin, I. I. 1999 How to define and calculate the degree of spin polarisation in ferromagnets. *Physical Review Letters* **83**, 1427–30.

Sun, J. Z. 1998 Thin-film trilayer manganate junctions. *Phil. Trans. Roy. Soc.* A **356**, 1693–711.

Ziese, M. 2000 Colossal magnetoresistance, half-metallicity and spin-electronics. *Phil. Trans. Roy. Soc.* A **358**, 137–50.

6
Polymer electronics

Ifor D. W. Samuel

School of Physics and Astronomy, University of St Andrews, Fife KY16 9SS, Scotland, UK

6.1 Introduction

Polymers, or plastics, are familiar materials found throughout everyday life. Polymer molecules consist of many repeat units and the enormous range of possible repeat units give an almost unlimited variety of polymeric materials. This gives great diversity and applications of polymers range from compact discs to bullet-proof vests and from cling film to car tyres. Biological polymers such as DNA, proteins and cellulose form the basis of life itself. One of the key reasons for the widespread use of man-made polymers is the ease with which they can be processed into almost any desired shape or form. They can be moulded into a particular shape, extruded to make tubes or films, or spun to make fibres. A further advantage of polymers is that their properties can be controlled by modifying their chemical structures. To date polymers have been used widely as structural materials (e.g. for furniture, computer housings, etc.), for packaging and as fabrics and films.

There are some remarkable polymers that can conduct electricity, leading to new directions in electronics and polymer science. They offer the prospect of materials that combine novel electronic properties with the ease of processing of polymers, whose properties can be tuned by chemical modification to give desired features. This new class of electronic materials means that we can now dream of giant flexible displays and electronic circuits made by printing. In this article we shall look at the remarkable

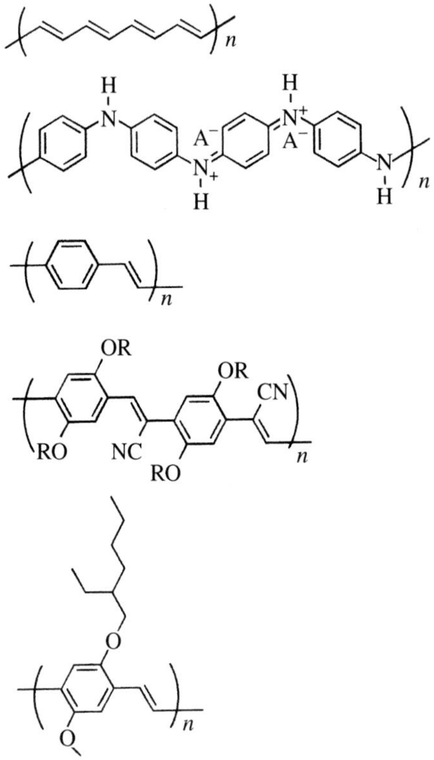

(*a*) polyacetylene

(*b*) polyaniline (emeraldine salt)

(*c*) PPV

(*d*) CN-PPV

(*e*) MEH-PPV

Figure 6.1. Chemical structures of a selection of conducting and semiconducting polymers. Polymers are long molecules consisting of many units and, by changing the repeat units, their properties can be controlled. The polymers shown are *conjugated*, which means that the backbone of the polymer consists of alternating single and double bonds, leading to their novel electronic properties. Polyacetylene is the simplest example of a conjugated polymer, whilst poly(*p*-phenylene vinylene), known as PPV, was used to make the first polymer light-emitting diodes.

developments that have taken place in this field, giving plastic transistors and light-emitting diodes, and then consider the ingredients that will fuse to give future innovations.

The vast majority of polymers are electrical insulators. However, a special class of polymers, known as conjugated polymers, exhibits semiconducting properties. 'Conjugated' means that the molecules have a backbone containing alternating double and single bonds. The simplest example is polyacetylene which is shown in Figure 6.1(a). Electrons in the double bonds occupy orbitals (called π-orbitals), which overlap along the polymer chain. These electrons are mobile and therefore provide a way of carrying current. There are two electrons in π-orbitals per repeat unit of the polymer chain and, because of the Pauli exclusion principle, this leads to a filled band and semiconducting electronic properties. The electrons can move easily along a polymer chain, but with difficulty between neighbouring polymer chains, so that these systems have a one-dimensional character. An enormous range of conjugated polymers has been synthesised and some examples are shown in Figure 6.1.

6.2 Conducting polymers

The initial interest in conjugated polymers in the 1970s was in their potential use as conducting materials. Conjugated polymers can be doped, in analogy to inorganic semiconductors, and very high levels of doping lead to many free charge carriers and hence high conductivity. It was suggested that it might be possible to make polymers with the conductivity of copper and the strength of steel. High conductivities were achieved in polyacetylene, but have not been exploited since the material is very sensitive to exposure to air.

A promising alternative is polyaniline (Figure 6.1(b)), which can be doped to give a conducting polymer. By choosing an appropriate dopant, polyaniline can be rendered both soluble and conducting, but with the important advantage of a much-improved stability in air. The conductivities are not high enough for these materials to replace metals, but they do open up a range of applications for which only modest conductivities are required. Examples include capacitor electrolytes, conductive coatings for electrostatic speakers, anti-static packaging, electromagnetic screening and stealth technology. In addition conducting polymers can be used to make transparent conductive coatings for display applications. In 2000, the Nobel prize for chemistry was awarded to Alan Heeger, Alan MacDiarmid and Hideki Shirakawa for the discovery and development of conductive polymers.

6.3 Semiconducting polymers

During the 1980s interest in the semiconducting properties of conjugated polymers developed. The way in which the polymers respond to light and the addition of charge was studied and the results were related to the one-dimensional character of the materials. Since conjugated polymers are semiconductors, they can be used to make semiconducting electronic devices, such as field-effect transistors, light-emitting diodes, photodiodes and solar cells.

6.4 Field-effect transistors

One of the first polymer devices was a field-effect transistor; its structure is shown in Figure 6.2. It consists of a heavily doped silicon substrate, with an oxide insulating layer onto which a film of polyacetylene is deposited and contacts are evaporated. The highly doped silicon acts as a metallic contact. In a transistor the current flowing between two contacts (the source and the drain) is controlled by the voltage applied to a third contact (the gate). In the polymer field-effect transistor, the gate voltage is applied to the doped silicon. The resulting electrical field changes the density of charge carriers in the polyacetylene layer, thereby controlling the current flowing through it from the source to the drain.

An attractive feature of this type of device is its relatively simple fab-

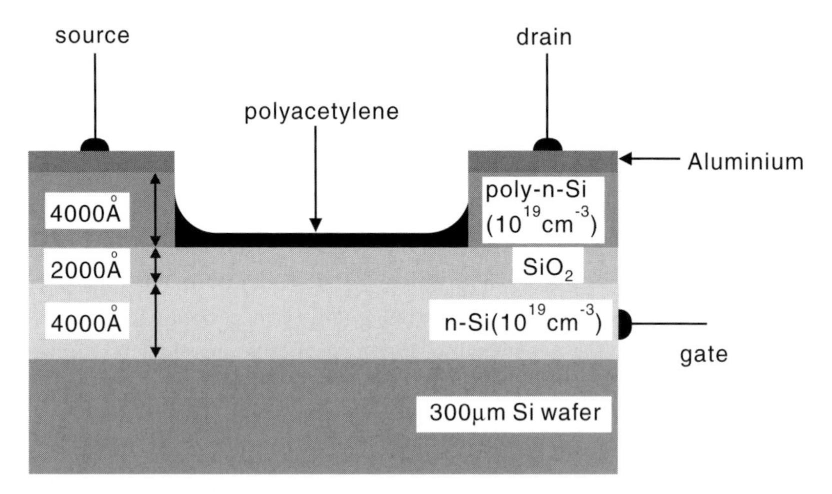

Figure 6.2. Thin films of conjugated polymers can be used to make a variety of semiconducting electronic devices such as transistors, light-emitting diodes and solar cells. One of the first of these devices was a polymer field-effect transistor. In a field-effect transistor, a voltage applied to one contact (the gate) controls the current flowing between two other contacts (the source and drain). The figure shows the cross section of one of these devices in which polyacetylene is used as the semiconducting polymer (adapted from Burroughes *et al.*, *Nature* **335**, 137 (1998)). Subsequent work has allowed flexible all-organic transistors to be made.

rication. All-organic transistors have also been made (i.e. transistors without a silicon substrate), giving even simpler fabrication. In addition to ease of fabrication, polymer electronic devices often have additional features not present in their inorganic counterparts. In the case of the polymer transistor, injection of charge forms states in the band gap that can absorb light, so that modulation of the gate voltage could be used to modulate the transmission of light through the device. Modulation of light is needed for encoding signals in optical telecommunications, although, in the case of a polymer transistor, the process is not fast enough to be useful. An alternative transistor involves using an organic material as the gate of the transistor. The organic layer can be doped by gases in the atmosphere, causing a large change in the current through the device when it is exposed to certain gases. This device can be used as a gas sensor and, by using a few such devices with different sensing layers connected to a neural network, it has been possible to make an 'electronic nose' (i.e. an electronic device capable of distinguishing between different smells).

6.5 Light-emitting diodes

In spite of the interesting developments outlined above, a decade ago, only a tiny fraction of the scientific community was aware of the possibility of polymers having semiconducting electronic properties. The situation has been transformed by the serendipitous discovery in Cambridge that polymers could be used to make light-emitting diodes (LEDs). The Cambridge group found that, when a voltage was applied to a conjugated polymer film in between two suitable contacts, the polymer emitted light. The polymer used was poly(*p*-phenylene vinylene), generally referred to as PPV (see Figure 6.1(c)) and it gave a yellow–green glow. Although the efficiency of light emission was very low, its enormous potential was quickly recognised and led to an explosion of interest in the field.

A simple polymer light-emitting diode is shown in Figure 6.3. It consists of a conjugated polymer film in between two electrodes. When a (DC) voltage is applied to the electrodes, negative charges (electrons) are injected from one contact and positive charges (holes) from the other. Some of the opposite charges injected meet up and emit light. The light leaves the device through the lower contact, which is made of the transparent conductive material indium tin oxide (ITO). The fabrication of such devices is simple: a thin film of the conjugated polymer is deposited onto an ITO-coated glass

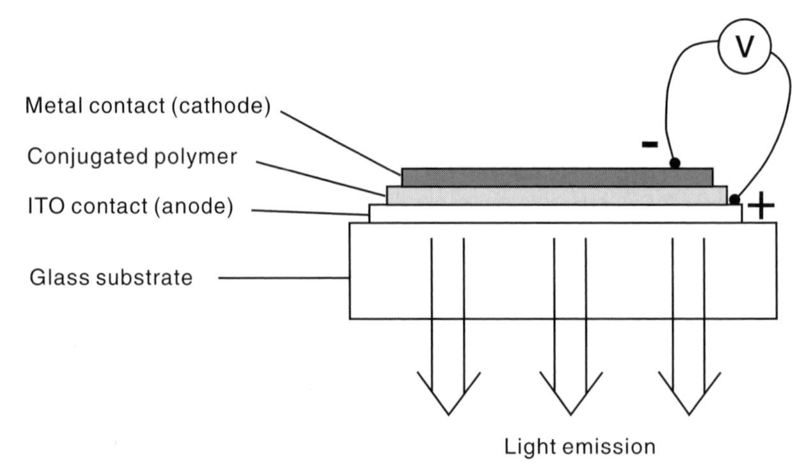

Figure 6.3. A cross sectional view of a simple polymer light-emitting diode. The device consists of a thin semiconducting polymer in between two contacts. When a voltage is applied, charges are injected into the polymer layer, leading to emission of light. The light leaves the device through a transparent contact, which is usually made of a material called indium tin oxide (ITO). This simple device forms the basis of a new display technology. It offers the prospect of flat displays that emit light, giving excellent visibility and viewing angle combined with a low operating voltage and relatively simple manufacture. Early applications are likely to include mobile-phone displays.

substrate by spin-coating and the top contact is then deposited by evaporation. The spin-coating step involves putting a few drops of polymer solution onto the substrate and then spinning it at high speed. The polymer solution flies off sideways, leaving a uniform thin film of the polymer on the substrate. Hence, although making a uniform thin film of thickness 100 nm might seem difficult, the flexibility of processing of polymers means that it can be achieved in a matter of seconds.

Polymer LEDs are very promising for display applications. They offer the prospect of displays that emit light and therefore have excellent visibility, contrast and viewing angle. They can be flat, operate at low voltage and are relatively simple to manufacture. Further advantages are their fast response (compared with liquid crystals), DC operation and robustness due to their all-solid-state construction. In the longer term this technology might be used for very-large-area and even flexible displays. The earliest applications are likely to be backlights for liquid-crystal displays and

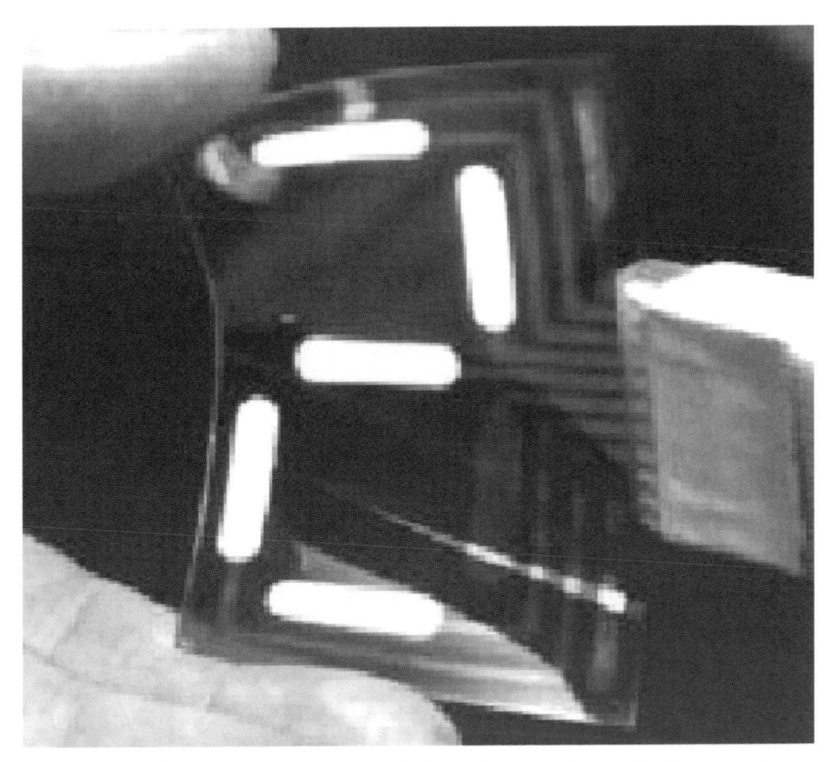

Figure 6.4. A photograph of a prototype light-emitting polymer display consisting of seven segments, each of which is a polymer LED (courtesy of Uniax Corporation). The light-emitting polymer has been deposited onto a flexible substrate to make a flexible display.

simple passively addressed displays such as for mobile phones. An example of a prototype polymer display is shown in Figure 6.4.

The operation of a polymer LED involves many steps, each with an associated efficiency, as depicted in Figure 6.5. First, opposite charges are injected from the two contacts into the polymer film. Some of these charges will meet and form an excited state called an exciton. Of the excitons formed, it is believed that three quarters will be in the triplet spin state (which does not emit light) and only one quarter in the singlet state (which can emit light). Only a fraction of the singlet excitons formed emits light. Finally, only part of the light generated in the polymer layer escapes from the device. The efficiency of light emission can be increased by increasing the efficiency of one or more of these processes.

(a)

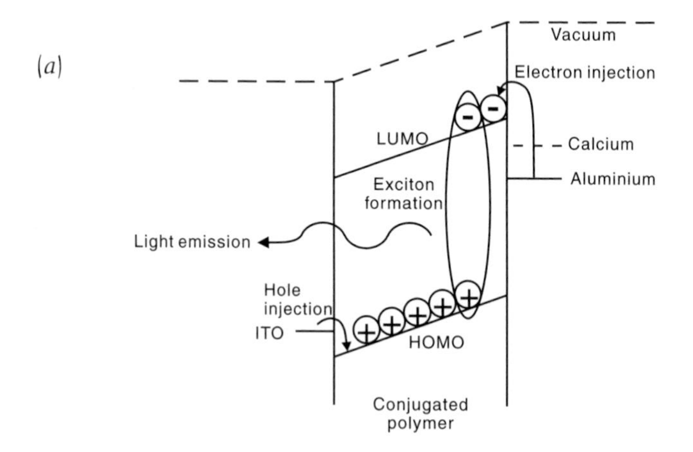

(b)

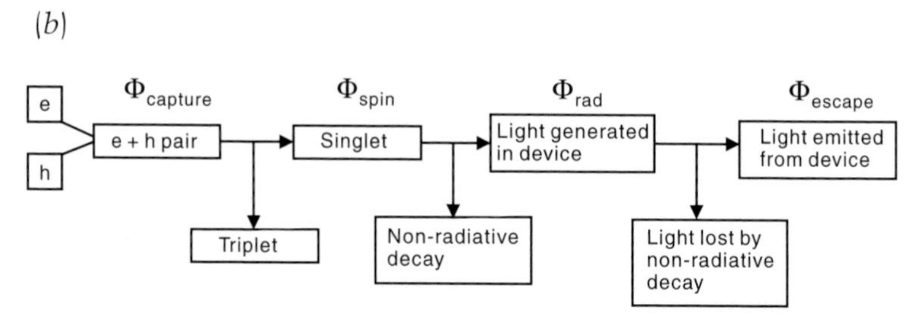

Figure 6.5. The operation of a polymer LED. (a) Energy levels of a LED under an applied voltage. HOMO denotes the highest occupied molecular orbital (or valence band). LUMO denotes the lowest unoccupied molecular orbital (or conduction band). The key point is that there is an energy barrier to the injection of charge from the contact into the polymer. (b) The sequence of events leading from charge injection to light emission. Positive and negative charges (electrons and holes) are injected from the contacts. A fraction $\Phi_{capture}$ of these meet up to form an excited state called an exciton. Of the excitons formed, only a fraction Φ_{spin} are singlets, the remainder being in the triplet spin state, which does not emit light. The singlet excitons decay by a combination of radiative and non-radiative processes. A fraction Φ_{rad} of the singlet excitons emits light. Finally this light has to leave the light-emitting diode. A fraction Φ_{escape} of the light generated escapes from the device, the remainder being trapped inside it. By optimising each step, the efficiency of the emission of light can be maximised.

The injection and transport of charge play a vital role in the operation of LEDs. In many early LEDs the injection of holes (positive charges) was much more efficient than that of electrons (negative charges). This is because, in a typical polymer such as PPV with ITO and aluminium contacts, there is a small energy barrier to the injection of holes at the ITO/polymer interface and a much larger barrier to the injection of electrons at the metal/polymer interface (see Figure 6.5(a)). In such devices, many holes pass straight through the structure without meeting an electron, leading to a low efficiency. The situation can be improved substantially by using an electron-injecting contact with a low work function such as calcium. A low work function means that the energy levels of the metal are such that the energy barrier to the injection of electrons is reduced, giving much more balanced injection of electrons and holes. Unfortunately, metals with low work functions are also very reactive, although the situation can be alleviated slightly by using alloys such as lithium–aluminium and magnesium–silver. An ingenious alternative is to modify the energy levels of the polymer to facilitate injection of electrons. This strategy was implemented by making the cyano-substituted PPV with increased electron affinity shown in Figure 6.1(d). Even if the injection of electrons and holes is balanced, their transport need not be. The mobility of electrons is generally much lower than that of holes, which leads to the opposite charges meeting up close to the metal electrode, which may then quench some of the luminescence.

The polymers used in LEDs must satisfy many criteria such as being highly luminescent, emitting light of the desired colour, forming good films by spin coating and having similar barriers to injection of electrons and holes. It is hard enough to make materials that satisfy these criteria and even harder when the requirement of similar mobilities for electrons and holes is added. An interesting alternative is to make a device with two or more layers of different polymers. One polymer can then be optimised for hole transport and the other for electron transport. This idea builds on the multilayer devices made from small organic molecules and inorganic semiconductors. A variety of electron-transporting polymers has been investigated and it was found that they give large improvements over similar devices without the electron-transporting layer. However, recent progress has given efficient devices without the need for a separate electron-transporting polymer. If this can be achieved for emission of red, green and blue light then simpler device structures may be acceptable.

Emission of light from conjugated polymers can also occur when they absorb light – a process known as photoluminescence. The light-emitting excited state is usually the same irrespective of whether it is generated electrically or optically, so that photoluminescence can give insight into the operation of LEDs. In particular, the efficiency of photoluminescence gives an estimate of the fraction of singlet excitons formed in an LED that will subsequently emit light. The value obtained is generally regarded as an upper limit because, as mentioned above, if excitons form close to the metal contact of an LED, there are additional quenching mechanisms. Improvements in the synthesis of polymers have increased photoluminescence quantum yields from a few per cent to approaching 100 per cent. The combination of materials and device structures that give balanced injection and transport of electrons and holes together with efficient radiative decay has led to remarkable progress. The efficiency of an LED can be expressed as the number of photons emitted divided by the number of charges passing through the device. This quantity is called the external quantum efficiency and has increased from 0.001 per cent to more than 10 per cent in only eight years. This improvement by four orders of magnitude in less than a decade represents outstanding progress. The efficiencies of the best devices now exceed those of incandescent light bulbs and many inorganic LEDs. These advances mean that it is now possible to contemplate polymer LEDs for lighting as well as display applications.

What are the prospects for even further increases in efficiency? In the best devices the efficiency of capture of opposite charges and the efficiency of emission of light from singlet excitons are nearly optimised. Substantial future improvements are therefore likely to be obtained by achieving emission of light from triplets or by increasing the fraction of light generated in the polymer layer that escapes from it. It has recently been shown that transfer of energy to a phosphorescent dye can be used to obtain emission of light from triplets, so this is a very promising future direction. There are also ways of increasing the amount of light escaping from a polymer film – for example by using wavelength-scale microstructure to achieve Bragg scattering of waveguided modes out of the film.

The stability of early polymer LEDs was very poor, so for a long time stability appeared to be a critical issue that could prevent the development of polymer LEDs. Fortunately the excellent progress in efficiency has been exceeded by advances extending the lifetime of devices. The operating lifetime of polymer LEDs has increased from minutes to more than 10000 h.

The main factors leading to this increase are improvements in materials and device-fabrication techniques together with effective encapsulation to keep out oxygen and water. There is still considerable scope for a fuller understanding of mechanisms leading to failure of devices, but current lifetimes are acceptable for many commercial applications. In parallel with research to give further improvements in efficiency and lifetime, much work is now being devoted to the challenge of making displays consisting of large numbers of polymer LEDs.

6.6 Other polymer devices

In LEDs, the recombination of charges gives emission of light. The reverse process, namely absorption of light leading to separation of charges, is also of great interest. It could form the basis of polymer solar cells and photodetectors. The requirements on materials and devices differ from those of LEDs: in an LED the aim is to make charges meet up, whereas in photovoltaic devices the aim is to separate charges. Absorption of light by conjugated polymers leads to the formation of a neutral excited state. In a photovoltaic device this excited state needs to be split into an electron and a hole and then the electrons and holes need to be transported to opposite contacts. A very successful way of achieving this is to make a blend of two polymers. It is well known that mixtures of polymers will separate into their different components. This leads to a interpenetrating network of filaments of the two polymers. By choosing materials with suitable energy levels so that an electron will move into one polymer and a hole will move into the other, separation of charges can be obtained with good efficiency. The separated charges can then be transported through each polymer to the electrodes. This structure can be used in two ways: a bias can be applied and it will then work as a photodiode; alternatively, in the absence of an applied bias, it can be used as a solar cell. Both applications would take advantage of the potential of making large-area devices from polymers. In the case of solar power, there is strong competition from other technologies, but the field of polymer photodetectors is more promising.

The development of a polymer laser is another target. So far there have been numerous reports of lasing and related phenomena, but they all use optical pumping – in other words, another laser is required to excite the polymer to make it lase. This is a first step towards a polymer laser diode, but there are many obstacles to overcome. In particular, losses associated with metal contacts, injection of charges and triplets will need to be overcome to make an electrically pumped laser a realistic proposition.

6.7 Other organic materials

Polymers are not the only organic semiconductors – there is a wide range of small organic molecules with semiconducting properties. Many of these molecules have been used to make excellent organic devices. For example, the oligomer sexithiophene can be used to make transistors with mobilities approaching those of amorphous silicon. LEDs based on small organic molecules have very good efficiencies and lifetimes and an impressive colour display has already been made by Pioneer in Japan. A key difference from the polymers is the way in which the small organic molecules are processed. Thin films of these materials are generally deposited by evaporation in a vacuum. This means that in the long-term polymers are likely to have advantages for low-cost manufacture (e.g. by printing) and large-area applications.

A potential alternative to either small molecules or polymers is to use conjugated dendrimers. The idea is to have a dendritic (highly branched) molecule. The core of the molecule can be chosen to have the desired electronic properties (e.g. colour) and connected by conjugated linkers to the surface groups which are selected to control the processing properties. Dendritic molecules have been used successfully as charge-transport layers and also light-emitting layers in LEDs. The potential of a family of dendrimers with various cores for tuning light-emitting properties has recently been demonstrated. The resulting molecules could be processed in solution, and gave light emission in the red, green and blue regions of the spectrum. The key advantage of conjugated dendrimers is that they can combine some of the advantages of small molecules with the solution processing of polymers.

6.8 Current and future developments

The future of polymer electronics will be shaped by innovations in materials and their processing, guided by understanding of structure–property relations and the operation of devices.

6.8.1 Designer materials

A vital part of the rapid development of polymer electronics to date has been the co-operation between chemists and physicists working to make better materials. There is no doubt that this trend will continue and that new developments will be governed by innovations regarding materials. There are two main approaches to improving the properties of semiconducting polymers. The first is largely a 'trial-and-error' approach in which a chemical modification is made and its effect then assessed to see whether the material is better or worse. The alternative is to attempt to design a material with desired properties and then synthesise it. It is probably fair to say that both approaches have contributed to the development so far of polymers for LEDs. Clearly the second approach is desirable, because of the enormous number of possible materials and the labour involved in making each one. However, it is possible only if there is a detailed understanding of how the properties of a material relate to its structure. This makes a detailed understanding of the physics and chemistry of conjugated polymers vital for their future development.

There are numerous questions that need to be answered and two examples are 'what is the nature of the light emitting excited state?' and 'what factors control charge transport in polymer LEDs?'. Information about the nature of the excited state can be obtained from the spectrum of light emission, the efficiency of light emission and the lifetime of the excitation once it has been formed. There has been a lively debate about this regarding the polymer PPV and its derivatives. For PPV prepared in Cambridge, the results show that the light-emitting excited state resides on a single polymer chain and it is called the singlet exciton. In contrast, in other conjugated polymers, including films of a cyano-substituted PPV with high photoluminescence efficiency, the emission of light is from an excited state, such as an excimer, which is delocalised over two or more neighbouring polymer chains. The result is significant because it shows that interaction between molecules must be considered in the design of highly luminescent molecules.

An understanding of the factors controlling transport of charge through polymer films is important for the development of materials for device applications. The contacts on a polymer electronic device are typically separated by 100 nm in the case of LEDs and solar cells and by tens of micrometres in the case of field-effect transistors. It is very unlikely that a single polymer molecule will connect the contacts: instead transport of charge from one contact to the other must involve many hops between neighbouring polymer chains. The ease with which these hops can occur will depend on the arrangement of the polymer chains and hence also on the interactions between polymer molecules.

The picture that is emerging is that it is not simply the chemical structure of the polymers that is important but also the way the molecules pack together – the morphology of the sample. The morphology of the sample is ultimately determined by the chemical composition of the polymer and the processing used. However, large parts of conjugated polymer films are often amorphous, making it difficult to obtain information about chain packing by structural techniques such as X-ray scattering, which are better suited to crystalline materials. Finding ways of processing conjugated polymers and even using processing to control properties will be important. For example, mobilities of charge carriers could be improved by increasing the order of the samples by using very regular molecular structures, stretch orientation or liquid-crystal phases.

There have been some notable successes of rational design of materials with desired properties. As mentioned earlier, one example is the use of a cyano-substitution on PPV to increase its electron affinity, thereby facilitating injection and transport of electrons. Another example is the use of the ethyl-hexyloxy side group to make the soluble polymer MEH-PPV (see Figure 6.1(e)). The asymmetrical substitution impedes crystallisation of the polymer, making it more soluble. A further example is tuning colour by making copolymers of two different materials to tune from the colour of one to the colour of the other. This trend will continue and in the future developments in relating the properties of conjugated polymers to their structures and morphologies will mean that more and more materials can be designed with desired properties.

However, it is not enough simply to know what structure would provide an ideal material; it is necessary to be able to make it. For example, we know that impurities that quench luminescence are bad for LEDs, but it is much harder to ensure that there are none. Developments in chemistry will be another vital part of the development of polymer electronics. Current improvements in synthesis are giving improvements in purity, control of end groups and control of relative molecular mass. By changing side groups, morphology and solubility can be controlled, allowing control of charge transport and processing. An important distinction from some areas of synthetic chemistry is that exceptionally high purity is needed. An impurity that quenches luminescence at a concentration of one part per million could cause a significant reduction in efficiency of an LED. One promising direction is a synthetic procedure known as ring-opening metathesis polymerisation because it gives control over the relative molecular mass, and a narrow distribution of relative molecular masses.

In the longer term, new approaches to the synthesis of polymeric electronic materials will develop. At present many organic LEDs have several layers, each separately deposited and adapted for a particular function (e.g. light emission, electron transport and hole transport). A preferable architecture would be to have single molecules extending from contact to contact and molecules consisting of different sections adapted to different functions. This idea is illustrated in Figure 6.6 for the example of a LED. The part of the molecule near the hole-injecting contact would have a chemical composition suited for transporting holes to a central section adapted for emission of light , whilst the other end of the molecule conveys

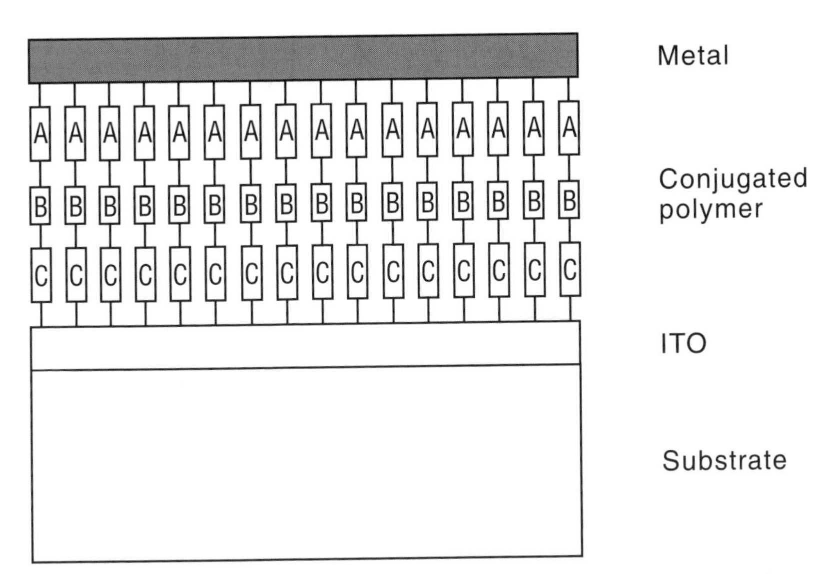

Figure 6.6. Innovations in materials science will continue to play a crucial role in the evolution of polymer light-emitting diodes. A possible future design of a LED in which ordered single molecules connect the contacts is shown in the figure. Each molecule combines electron-transporting (A), light-emitting (B) and hole-transporting (C) functions.

electrons to the light-emitting region. Ideally the molecules would be highly ordered to allow efficient charge transport and give a film with high physical robustness. A further development of this idea would be to incorporate segments of conducting polymers on each end of the molecule instead of conducting contacts. Even better would be if each section of the molecule were designed to associate with similar sections of other molecules. This could mean that, if the molecules were mixed together in solution, then they would associate to form a light-emitting polymer layer. Such an architecture would give endless possibilities of tuning properties – by changing repeat units, mixing different repeat units, changing the length of segments, or even having a continuous gradation of properties rather than discrete segments.

Whilst the idea above is at present a dream, some of the ingredients required are already discernible. The rapidly developing area of supramolecular chemistry is concerned with assemblies of molecules and interactions between molecules, linking to the theme of morphology discussed earlier. There have been attempts to make molecular wires and even molecular diodes. A promising approach is the use of 'self-assembly' techniques in which a structure is built up by depositing successive layers of molecules in such a way that there is a chemical bond between each layer and the next.

The control of order is widely recognised to be important. A possible future approach is to use a template such as a system of nanotubes formed in a liquid-crystalline mesophase of silica. These could be used to make an array of straight conjugated polymer molecules. Alternatively, a suitable rigid conjugated polymer that could form such a mesophase itself without the need for a silica template might be developed.

Once again we return to the theme of needing ways of making molecules suitable for the ideas outlined above. We would like a way of making any desired sequence of monomer units to give a polymer with particular properties. This may sound far-fetched, but it is already prevalent in the area of molecular biology, in which desired sequences of nucleotides and amino acids are commercially available. The same facility would revolutionise polymer science. The seeds of such an approach are emerging: the solid-phase synthesis of thiophene and other oligomers using an approach inspired by Merrifield has recently been reported. The impact of precise control over structures is evident in the field of inorganic semiconductors, in which molecular beam epitaxy and chemical vapour deposition have given exquisite control over properties of materials.

6.8.2 Flexible electronics

A very exciting aspect of polymer electronics is the possibility of building flexible electronic circuits by depositing polymer layers on suitable flexible substrates. Perhaps the first question to consider in this context is that of whether anybody would want flexible electronics. In some ways this is a test of imagination. In the context of displays, one significant use of flexibility would be to allow a particular shape of display to be realised. For example some displays (e.g. car dashboards) are curved, allowing optimal use of space and helping to keep off stray light. For some of these applications there is no need to change the shape of the display once it has been fitted, but flexibility would provide a convenient way of realising a particular shape. However, there could also be uses for flexible displays – for example, they could easily be stored by folding (or rolling) them up when they are not in use. In the context of electronics, smart cards that contain a simple chip and electronic bar-codes are growing in use. It is desirable for such devices to be flexible and polymer field-effect transistors provide a possible way of achieving this.

However, there are two additional implications of flexibility that go far beyond the uses described above. The first is that flexibility confers robustness. An ability to bend means that brittle fracture can be avoided. This is relevant to all applications, but especially those involving very-large-area devices. For example, a display panel of dimensions of a few metres would be extremely vulnerable to breakage unless it were flexible or had a very cumbersome support. The other important implication is that flexibility dramatically enhances manufacturability. It makes available the possibility of having a process that begins with a drum of substrate that is then coated with appropriate polymer layers and then finally sliced into polymer electronic devices in a single continuous process.

An excellent demonstration of the feasibility of flexible polymer LEDs was made in Santa Barbara and this has subsequently been developed into the device shown in Figure 6.4. These devices are made on a poly(ethylene terephthalate) substrate (as used for transparencies for overhead projectors) coated with conducting polyaniline as a hole-injecting contact. A semiconducting polymer layer is deposited on top of this and the device is completed by a top contact of calcium. As can be seen in Figure 6.4, the device can be bent and still operates perfectly. Philips has demonstrated flexible electronic tags. Flexibility could be useful in the future for lighting applications. Existing light sources are generally very bright over a very small area (e.g. the filament of an incandescent lamp). A very-large-area light-emitting polymer device could be used to give a very different and perhaps more restful illumination in which low brightness over a large area is used. Flexibility could also be exploited to make light-emitting clothing for safety or fashion applications.

A major disadvantage of using polymer substrates is that they are far more permeable to air and water than is glass, which is more commonly used as a substrate. This is a problem because oxygen and water cause degradation of polymer devices. There are two main approaches to dealing with this problem: the first is to find or design substrate materials with improved barrier properties; the second is to develop conjugated polymers that are more resistant to air and water.

6.8.3 Printable electronics

Advances in polymer processing will play an important role in the development of polymer electronics. Printing is the most exciting area because it really exploits the favourable polymer-processing properties mentioned in the introduction. The greatest advantage of printing is its potential for mass production and patterning. It could be used to define contacts for pixellated displays or simply to define the shape of a light-emitting company logo. In addition it is an additive process and therefore minimises wastage of material. Some forms of printing would also be well suited to rapid prototyping or small-scale customised production, whilst other forms could allow mass production.

Progress in printing polymer electronics is very encouraging. Ink-jet printing of polymer LEDs has been demonstrated and it has also been shown that FETs can be printed with respectable charge carrier mobilities. Even more impressive is the demonstration of a prototype miniature television screen 50 mm square and 2 mm thick. It results from the alliance between Cambridge Display Technology and Seiko-Epson and was made by ink-jet printing of polymers onto polysilicon thin-film transistors. The prototype is yellow–green and the companies are now working towards a full-colour version of the display.

In future we can expect to see a dramatic increase in the use of printing for the fabrication of polymer electronic circuits. One area of particular interest would be to combine field-effect transistors and LEDs to make an actively addressed polymer display. There are many technical challenges to achieving this, but the possibility of combining these devices in structures made by spin-coating has recently been demonstrated.

As printing develops, we can expect to see it develop towards higher and higher speed. Basic lithographic printing presses can achieve 10000 impressions per hour. A significant development has been a conducting ink compatible with lithographic printing. The ink consists of conducting metal particles in a polymer matrix. It allows conductive circuits to be printed at very high speed. The conventional process for making 'printed' circuits involves a photographic exposure followed by etching, which generates acid waste. The lithographic approach to printing is dramatically faster and cleaner. If it can be used to print polymer electronic devices, it would allow the fabrication of electronics at unprecedented speed and low cost. This could transform the range of applications of electronics.

6.9 Outlook

The impact of materials is so great that eras of history are named after them. In the modern age, improvements in materials have transformed our lives with developments from jet engines to skyscrapers. In this article we have seen how conducting and semiconducting polymers are leading to new directions in electronics and polymer science. The development of polymer LEDs from discovery to the point of commercial manufacture in a decade is remarkably fast compared with other areas of semiconductor electronics. Simple polymer displays such as passively addressed displays for mobile phones and backlights for liquid-crystal displays are poised to enter the marketplace. The subsequent development of the field will depend as much on commercial as on technical issues. Assuming that the early displays are successful, we can expect to see colour displays with increasing information content.

In the longer term, applications that make the maximum use of polymer processing properties are very significant. The scope for flexible electronics and very-large-area displays is very exciting. Even more significant is the possibility of printing electronics. Beyond that it is possible to envisage combining many devices onto a single sheet of plastic so that display, logic, light detection and even power generation are all integrated. The technical challenges should not be underestimated, but neither should the potential rewards.

The potential of this field is so great that it has been possible only to select a few aspects of it in this article. Numerous opportunities exist at the boundaries with other research fields. For example, there is a healthy symbiosis with the field of quantum optics: luminescent polymers can be used to make novel photonic structures and, conversely, photonic structures can be used to control the properties of polymers, including lasing. There is also the potential for hybrid devices combining organic and inorganic materials – the compatibility of semiconducting polymers with silicon is an attractive feature. Equally exciting is the scope for polymer electronics to inspire new chemistry and for new physics to emerge from the new materials and devices.

6.10 Further reading

Bradley, D. D. C., Friend, R. H., and Holmes, A. B. (editors) 1997 Electronics with molecular materials: from synthesis to device. *Phil. Trans. Roy. Soc.* A **355**, 691–842.

de Leeuw, D; Kido, J. and other articles. 1999 *Physics World* **12**, March.

Friend, R. H. *et al.* 1999 Electroluminescence in conjugated polymers. *Nature* **397**, 121–7.

Friend, R. H., Burroughes, J., and Shimoda, T. 1999 Polymer diodes. *Physics World* **12**, 35–40.

Gustafsson, G., Cao, Y., Treacy, G., Klavetter, F., Colaneri, N., and Heeger, A. 1992 Flexible light-emitting-diodes made from soluble conducting polymers. *Nature* **357**, 477–9.

http://www.cdtltd.co.uk; http://www.uniax.com

Samuel, I. D. W. 2000 Polymer electronics. *Phil. Trans. Roy. Soc.* A **358**, 193–210.

Yam, P. 1995 Plastics get wired. *Scientific American* July, 74–9.

See also: http://www.nobel.se/chemistry/laureates/2000/

7
Quantum-enhanced information processing

Michele Mosca[1,3], Richard Jozsa[2], Andrew Steane[1] and Artur Ekert[1]

[1] Centre for Quantum Computation, Clarendon Laboratory, Department of Physics, University of Oxford, Parks Road, Oxford OX1 3PU, UK
[2] Department of Computer Science, University of Bristol, Woodland Road, Bristol BS8 1UB, UK
[3] Department of Combinatorics and Optimization, University of Waterloo, Waterloo, ON, Canada N2L 3G1

7.1 Quantum information: bits and qubits

The handling of information is becoming an increasingly important part of everyday life. Anyone who has enjoyed listening to music on a CD, watching a video or browsing the Internet, or has used an automatic bank teller machine, has benefited from the recent explosion of developments in processing, storage and transmission of information.

A given piece of information may be expressed in many different forms. For example a cake recipe may be written in English or in Chinese. It may be spoken or stored in a computer memory coded as a sequence of 0s and 1s. Note that written words are arrangements of ink molecules on paper, spoken words are fluctuations in air pressure and a computer memory can be constructed from various kinds of electromagnetic components. In fact, all forms of information have a fundamental common feature: they all use physical objects to represent the information and processing is always performed by physical means. It follows that the possibilities of and limitations on storage and processing of information are ultimately dictated not by mathematical constructions, but by the laws of physics.

To use a physical system for information storage we must first identify and label a number of its distinguishable states. The basic unit of information is the *bit* (a contraction of 'binary digit') which is represented by any physical system with just two distinguishable states, labelled 0 and 1.

Any information can be represented by suitable sequences of bits. For example, the twenty-six letters of the Roman alphabet may be unambiguously coded using some twenty-six of the thirty-two possible five-bit strings (whereas the sixteen possible four-bit strings do not suffice) and any written text is then represented as a sequence of bits.

In a digital electronic computer two levels of voltage are used to represent the bit values 0 and 1. One bit of information can also be encoded in two different polarisations of a photon or in two different electronic states of an atom. In the latter cases the physical system is governed by the laws of quantum physics and is not well described by the formalism of classical physics.

This article will discuss the consequences of the fact that these fundamental differences between quantum and classical physics may be exploited to give rise to novel methods of storage and processing of information, which in principle go well beyond the capabilities of current technology that is based on classical representations of information.

Consider a bit coded in a quantum system such as the polarisation of a photon. It is customary in quantum physics to label states using a curious notation of a half-pointed bracket enclosing a label. The two distinguishable states representing the bit values 0 and 1 are written as $|0\rangle$ and $|1\rangle$. This so-called *ket* notation was introduced by P. Dirac in the 1920s to facilitate mathematical calculations. For us its odd appearance will serve as a constant reminder of the weirdness of quantum phenomena! According to the laws of quantum mechanics the system can also be prepared in a *coherent superposition* of the two basic states. This is mathematically written as $|\Psi\rangle = a|0\rangle + b|1\rangle$ where Ψ is the label of the superposed state and a and b are complex numbers. In such a state the system is interpreted as being simultaneously in *both* states $|0\rangle$ and $|1\rangle$ (to varying degrees depending on the values of a and b). Under physical evolution the component parts $|0\rangle$ and $|1\rangle$ evolve separately and the system indeed behaves in a non-classical way as though it were simultaneously in states $|0\rangle$ and $|1\rangle$. Any such quantum system that can encode the basic bit values as well as any possible superposition is called a *qubit* (pronounced 'queue bit').

An especially important feature of the quantum behaviour of qubits arises when we consider a string of several bits or qubits in a row. Consider first a row of two bits. There are four possible states, 00, 01, 10, and 11. If these are coded in a quantum system (i.e. we have two qubits) then, in addition to the basic states $|0\rangle|0\rangle$, $|0\rangle|1\rangle$, $|1\rangle|0\rangle$ and $|1\rangle|1\rangle$, we now also have general superpositions $|\Phi\rangle = a|0\rangle|0\rangle + b|0\rangle|1\rangle + c|1\rangle|0\rangle + d|1\rangle|1\rangle$. Thus two qubits may simultaneously represent all four possible 2-bit strings (in a particular quantum way depending on the coefficients a, b, c and d). More generally, for n classical bits there are 2^n possibilities but any single state of n classical bits can be described by a bit string of length n. In contrast, n qubits may simultaneously include all 2^n possibilities in superposition. In this sense n qubits are able to embody vastly more – in fact exponentially more – information than n classical bits. The gap between n (classical) and 2^n (quantum) grows very rapidly with increasing n. This exponentially enhanced richness of multi-qubit states for representing information has profound consequences for information processing and communication, as described in the following sections.

Another fundamental non-classical feature of states of two or more qubits is the phenomenon of *quantum entanglement*. A general superposition state of two qubits (such as $|\Phi\rangle$ above) has the property that each separate qubit cannot be assigned a separate state of its own. For example the 2-qubit state

$$|\chi\rangle = \frac{1}{\sqrt{2}}|0\rangle|0\rangle + \frac{1}{\sqrt{2}}|1\rangle|1\rangle$$

cannot be expressed as a juxtaposition of 1-qubit states $(a|0\rangle + b|1\rangle)(c|0\rangle + d|1\rangle)$. In contrast, in any state of two classical bits (e.g. 01), each bit separately has a well-defined value and the whole is just the juxtaposition of the parts. In the quantum case the information of the state does not reside locally in the separate qubits but is also distributed non-locally in a rich variety of possibilities of correlations between basic superposition components.

Only the totality of all qubits together has a well-defined state. The qubits are thus said to be *entangled*. Note that classical bits may also exhibit correlations. For example, two bits might be prepared either as 00 or 11 but we do not know which. Then, on examining the value of the first bit, we immediately learn the value of the second bit too – they are perfectly correlated. However, the quantum correlations involved in entanglement are far richer. It can be shown, for example, that the kinds of correlated behaviour exhibited by the entangled state $|\chi\rangle$ above cannot be reproduced by any model with just pre-assigned classical bit correlations.

A particularly interesting situation arises when entangled qubits are separated in space. This is not unusual in nature. For example, if an electron in a calcium ion is excited to a higher energy level and then allowed to fall back to its lower level, the excess energy is emitted in the form of two photons, flying off in opposite directions. These photons are physically entangled in a state very similar to $|\chi\rangle$ above. The entanglement of the photons is independent of their spatial separation. They exhibit their peculiar quantum correlations even though there is no tangible physical connection between them. No physical process in the space between the photons can affect their entanglement. The correlation properties of such spatially separated entangled systems may be exploited for a variety of novel communication tasks, including the possibility of perfectly secure classical communication (which is impossible with classical bits) and the process of 'quantum teleportation' which will be described in later sections.

So far we have discussed how information can be *embodied* in the state of a physical system. An important dual aspect of this physics of information is the question of how the information can be *accessed* or *read out*. In terms of physics, this is the question of what kinds of *measurements* are possible. Again we find a dramatic difference between classical and quantum physics. Information in classical physics can in principle always be read out completely and perfectly. In contrast, the laws of quantum physics (especially the uncertainty principle) imply that any attempt to read the information embodied in a quantum state will irretrievably disturb the state. Only a small amount of the potentially vast information content can be read out and most of it must remain inaccessible!

The full 'unknowable' information embodied in the identity of a quantum state is called *quantum information*. It has many peculiar properties. For example, we are all familiar with the idea of copying classical information – a cake recipe can be photocopied, copied out by hand or read out over the telephone, giving two or more copies of the information. In 1982, in a famous 'no-cloning' theorem, W. Wootters and W. Zurek showed that quantum information cannot be copied! For example, if quantum information is communicated from A to B then the information is necessarily completely destroyed at A as it appears at B and no record of its identity can remain at A. The inaccessibility and non-clonability of quantum information may appear at first sight as entirely negative features but these strange properties may be usefully exploited! For example they can provide a means of perfectly secure communication, as we will elaborate later. Roughly speaking, any attempt to eavesdrop on the information must leave its imprint on the quantum state, which could later be detected by the legitimate communicating parties.

As any physical system evolves in time, the identity of its state changes. Thus quantum physical evolution can be naturally viewed as the processing of quantum information. The laws of physics allow us to predict and *calculate* the changing identity of the state. In 1982 R. Feynman made a remarkable and profound observation: if we calculate or 'simulate' the quantum evolution on any standard computer then the amount of computational effort involved generally grows enormously as time passes. With each successive second of temporal evolution of the actual quantum physical system, the amount of computational effort needed to simulate it will grow so rapidly that soon the temporal and spatial requirements for the simulation will exceed all available resources. In fact the resource requirements grow 'exponentially' – an important notion that will be elaborated in the next section.

On collecting the above ideas together we arrive at a bizarre picture of the quantum world: in ordinary temporal evolution nature processes quantum information at an astonishing rate that cannot be matched by any conventional computer simulation, yet when the processing is finished, most of the information is kept hidden and inaccessible to being read!

A quantum computer is any physical device that exploits the greatly enhanced information-processing power of quantum evolution for computational purposes. The very restricted accessibility of the processed quantum information provides a severe limitation on our ability to exploit the enhanced computing power. However, it does not annul it! As a basic illustrative example, suppose that we have a quantum computer programmed to compute a function f. The computer evolves the labelled input state $|x\rangle|0\rangle$ to the labelled output state $|x\rangle|f(x)\rangle$. (Here the second ket, initially set to $|0\rangle$, is the output register for the value of the function.) Now we may prepare the input register in an equal superposition $\Sigma_x|x\rangle$ of *all* possible input values x. Running the computer then yields the output state

$$|f\rangle = \sum_x |x\rangle|f(x)\rangle,$$

i.e. by evaluating the function *once* we evaluate *all* function values $f(x)$ in superposition. This process is called computation by quantum parallelism. The quantum information of the state $|f\rangle$ includes information on all of the $f(x)$ values but because of its inherent inaccessibility we are unable to read them out. Nevertheless, small amounts of 'global' information – relating to all function values simultaneously – may be read out and this information could still require a vast (exponential) amount of computing effort to obtain on a conventional classical computer. For example, we may wish to discern simple patterns in the list of values such as periodicity if the function is periodic (see the next section). In this way we may successfully exploit the greatly enhanced information-processing power of quantum evolution despite its inaccessibility. In the next section we will elaborate on the idea of computational complexity and give some interesting fundamental applications of quantum computation.

7.2 Computational complexity and quantum computers

A computer is a piece of hardware that runs according to a program, or *algorithm*, that we specify depending on the task we wish to perform. The computer carries out the algorithm on a given input, producing the desired output. The hardware could be an old Commodore 64, for example, and the algorithm could be a simple spell-checking program. This program takes a file of text as input and outputs a list of the misspelt words. Modern computers can take human-voice input and translate it into voltages representing 0s and 1s, which in turn represent the sound patterns. These 0s and 1s are processed via the computer into other 0s and 1s representing the words corresponding to those sound patterns. These voltages are translated into various colours of light emanating from a colour monitor, which visually displays the text corresponding to the spoken words. All of these translations and manipulations are controlled by hardware running according to algorithms in response to inputs.

The difficulty, or *computational complexity*, of a task is the amount of resources, such as time, space and energy, necessary to perform it. The computational complexity of a task of course depends on the size of the particular input. For example, the number of steps necessary to implement spell-checking of a document with n bytes of text is proportional to n. Another example of a computational problem is multiplication. The input is a pair of n-digit numbers and the output is the product of these two numbers. The simple multiplication technique taught in primary school is an algorithm that uses roughly n^2 steps to compute the product. The reverse task, factorisation, is to take an n-digit number (let us assume that it is not prime) and to output two smaller numbers that are multiplied together to produce that number. The best-known rigorous classical algorithm for performing this task uses over $10^{\sqrt{n}}$ steps. The number $10^{\sqrt{n}}$ gets astronomically larger than n or n^2 as n grows, so performing this task even for n as small as 200 is beyond the computing power available on the earth today. When n is 400, $10^{\sqrt{n}}$ is 10^{20}. Even one thousand computers running at 1000 MHz would take three years to perform 10^{20} operations. When n is 900, it would take one million computers, running at 1000 GHz, 10000 years to perform this many operations. As n grows only slightly, the difficulty of the problem quickly grows exponentially larger and quickly becomes intractable on any conceivable (classical) computing device. Multiplying numbers or spell-checking documents of these sizes, however, can be done in a fraction of a second. The latter two tasks are considered *tractable*, whereas the factoring problem is considered *intractable*. Another example of a problem that is considered intractable is that of deciding, for a given map of n countries, whether it can be *properly* coloured using only three colours, that is, whether it can be coloured in such a way that no adjacent countries are coloured the same. For many maps the answer is simple, but there are huge families of maps for which this problem is believed to be very hard and the best known algorithms require roughly 3^n steps.

The exact number of steps necessary depends on the details of implementation, which do not seem to affect the difficulty of the problems qualitatively. The reason why we do not worry about the details of implementation is that we believe that any 'reasonable' computing device can efficiently simulate any other 'reasonable' computing device, digital or analogue, mechanical, electronic or optical. Quantum computers, if they are indeed 'reasonable' computing devices (see Section 7.5), pose a serious problem for this belief since we believe that they cannot be efficiently simulated by any classical device. This is the essential content of Feynman's observation mentioned in the previous section. There are several tasks that are known to be tractable on a quantum computer yet are very strongly believed to be intractable on any classical computer. The most famous example is the above-mentioned factorisation problem. In 1994, P. Shor discovered a quantum algorithm for factoring an n-digit number, which runs for less than n^3 steps (which is feasible, unlike the $10^{\sqrt{n}}$ steps used by classical algorithms). See Figure 7.1 for an illustration.

Quantum computers also help with problems of less structure. Consider the problem we mentioned earlier, that of properly colouring a given map using only three colours. This problem has the property that we can easily check solutions (since there are fewer than n^2 pairs of adjacent countries to check), but deciding whether any solution exists can be very

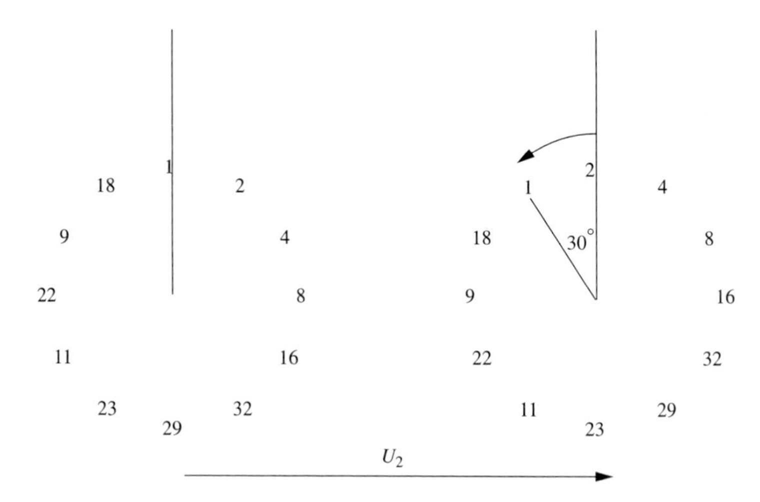

Figure 7.1. Quantum factoring. The problem of factoring the number N is closely related to studying the sequence 1, 2 mod N, 2^2 mod N,..., 2^x mod N, ..., where y mod N is the remainder when y is divided by N (e.g. 79 mod 35=9). This sequence will eventually start to repeat and cycle through the same numbers. To factorise N it suffices to find the period of sequences like this (replacing 2 by another number if necessary). That is, we wish to find the smallest positive number r such that the sequence repeats or cycles after every r steps. We do not care about the actual values in the sequence, apart from the fact that they cycle. The difficulty of this problem is very important since most of the encryption methods used by industry and on the Internet (e.g. to protect your PIN or credit-card number) rely on the intractiblity of this problem. However, quantum computers have an edge over classical computers in studying such global properties or patterns. Probing just a few values of this sequence gives little chance of finding the period: this approach requires looking at roughly $\sqrt{N}$ elements of the sequence. Quantum computers, however, can be in a state containing all the elements of this sequence. A physical realisation of this sequence, when it is multiplied by 2, will shift and will cycle back to its original state after r shifts. An object that cycles after r steps must be in a sense rotating at a rate of k/r cycles per step for some integer k. A physical realisation of this whole sequence, which a quantum computer can efficiently create, will rotate at such a rate, so a quantum computer can study this rate of rotation by looking at superpositions of elements in the sequence rather than at individual entries. The illustrated example uses N=35 and we seek the period r = 12. The operation U_2 (multiplication by 2 mod 35) sends 1 to 2, 2 to 4,..., 9 to 18 and 18 to 1. Individually, these bits of information give little clue that 2^{12} mod 35 is 1. Multiplying the above combination of all the powers of 2, however, will rotate the combination at a rate of 1/12 revolutions (30°) per multiplication. This

difficult (for some maps, no significantly better method is known than simply trying all 3^n possible colourings to see whether any are proper). Let us denote by *Verify_Colouring(c)* the function which takes as input such a colouring, c, and outputs 1 if it is proper and 0 otherwise. A quantum computer can evaluate *Verify_Colouring* on a superposition of all the possible colourings c but we cannot force it to observe a solution with *Verify_Colouring(c)*$=1$ using only one evaluation of f. The best known classical approaches to this problem require about 3^n evaluations of *Verify_Colouring* in order to decide whether a proper colouring exists. However, using a quantum search algorithm developed by L. Grover in 1996 (illustrated in Figure 7.2), we can cleverly make the many superposed components interfere in such a way that $\sqrt{3^n}$ evaluations suffice.

In summary, some problems, such as factorisation, have an algebraic structure that quantum computers can exploit to a much greater extent than can any known classical algorithm. These problems that were once felt to be intrinsically hard, that is, not efficiently solvable by any reasonable computing device, we now know can be solved in very few steps on a quantum computer. Also, simple searching algorithms can be speeded up by a square-root factor.

7.3 Communication and security of information

The previous section illustrates how exploiting the quantum nature of information can have an enormous impact on the computational complexity of many problems. However, information is a valuable resource that we may wish to share with our friends, or perhaps keep out of the hands of our foes or competitors. It is natural to ask whether the quantum nature of information changes the rules of the game in the communication and security of information. The answer is YES, quantum physics has a

rate can be approximated by a simple quantum algorithm and the denominator, 12, can be extracted. So now we know that $2^{12} \bmod 35 = 1$ and this information can be used to factor 35. The factors of 35 are no big secret, but try finding those of 27 997 833 911 221 327 870 829 467 638 722 601 621 070 446 786 955 428 537 560 009 929 326 128 400 107 609 345 671 052 955 360 856 061 822 351 910 951 365 788 637 105 954 482 006 576 775 098 580 557 613 579 098 734 950 144 178 863 178 946 295 187 237 869 221 823 983, the RSA 200-digit challenge (RSA is a widely used encryption procedure). A quantum computer could.

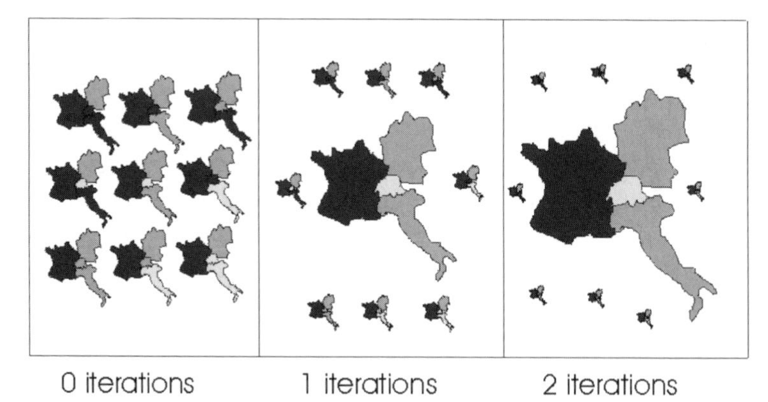

Figure 7.2. Quantum searching. One problem to which we can apply the quantum searching algorithm is that of colouring maps. Similar problems include the allocation of frequencies in mobile-phone networks so that neighbouring transmitters do not emit interfering signals. In this example of the map-colouring problem, we seek to colour the four countries using the three colours BLUE (shown the darkest), YELLOW (shown the lightest) and GREEN so that no adjacent countries are coloured the same. By simply preparing a uniformly weighted combination of all possible colourings (we assume, without loss of generality, that France is coloured blue and Germany is green, leaving $N=9$ possible colourings to consider for Italy and Switzerland) we get the unique proper colouring with probability 1/9. The first iteration of the quantum search procedure increases the probability of observing the proper colouring to roughly 73 per cent and a second iteration gives us the proper colouring with probability over 98 per cent. This is a good time to stop and observe a colouring. Of course, for such a small map the answer is obvious; however, for even as few as $n=50$ countries this problem can be a nightmare. Exhaustively trying all $N=3^{50}$ (this is over 10^{23}!) colourings is not feasible, whereas the number $\sqrt{N}=3^{25}$ (less than 10^{12}) of steps required by the quantum algorithm is within the realm of possibility.

dramatic effect. We will describe three examples: quantum teleportation, quantum key distribution and quantum complexity of communication.

As mentioned earlier, any interaction with a quantum system that extracts information about its state will disturb the system. In fact, any non-trivial interaction between the quantum state and its environment will alter the state. This means that quantum information is extremely delicate and, to remain fully intact, it must not interact with its environment. This makes storing and transporting quantum information a very challenging task. The error-correcting codes that we will discuss later employ entanglement and redundancy as means of keeping quantum information intact and resistant to interactions with the environment.

Now suppose that two people, Alice and Bob, are separated in space and wish to communicate some quantum information. A powerful technique, known as *quantum teleportation*, developed by C. Bennett, G. Brassard, C. Crépeau, R. Jozsa, A. Peres and W. Wootters in 1993, allows Alice and Bob to communicate quantum information by sending only a small amount of classical information. The advantage is that 'classical' information is very robust and much less sensitive to interaction with the environment. To achieve this task, Alice and Bob must also be in possession of some shared entanglement. More precisely, at some point in the past Bob created the entangled pair of qubits

$$|\chi\rangle = \frac{1}{\sqrt{2}}|0\rangle|0\rangle + \frac{1}{\sqrt{2}}|1\rangle|1\rangle$$

and carefully gave one half to Alice. With this resource, at any time in the future Alice can send Bob a quantum state $|\Psi\rangle$ by performing a special quantum measurement on $|\Psi\rangle$ and her share of $|\chi\rangle$ and sending Bob the result of this measurement classically. Bob can then reproduce the state $|\Psi\rangle$ and Alice no longer has her copy of the state. The protocol is described in Figure 7.3. The quantum teleportation of states of light has been realised in several laboratories around the world.

We next turn to the issue of security of communication. It is interesting to note that a commonly used method of secure communication – the method of public-key cryptography – relies for its security on the computational intractability of certain computational tasks such as factoring large numbers. As discussed in the previous section, the computing power of quantum processes can be used to break such ciphers and render them insecure! However, we will now see that further quantum effects – the uncertainty principle and the inaccessibility of quantum information – may be exploited to provide new methods of communication that *are* unconditionally secure. They do not rely on any unproven assumptions about computational intractability.

The sensitivity of quantum information to interaction with its environment might seem like nothing but an inconvenience, but it is a very useful tool in the art of secret communication. One very useful primitive in cryptography is the distribution of a common secret key between Alice and Bob. The key itself is random, that is, it contains no valuable information. What is valuable is the fact that Alice and Bob have the same key (100

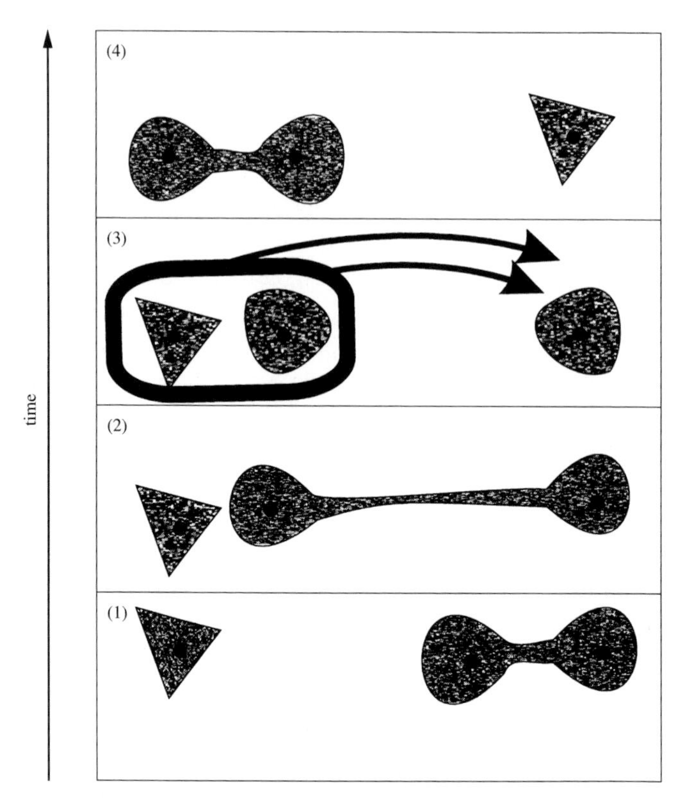

Figure 7.3. Quantum teleportation. Alice and Bob can communicate quantum information by sharing entanglement and sending classical bits of information. In step 1 Bob prepares two qubits in the state $|\chi\rangle = \frac{1}{\sqrt{2}}|0\rangle|0\rangle + \frac{1}{\sqrt{2}}|1\rangle|1\rangle$. He then sends one of the two qubits to Alice (step 2), who has a special qubit $|\Psi\rangle$ she wishes to send Bob at some point in the future (she can in fact decide at any point before teleportation what state $|\Psi\rangle$ she wishes to send). Although the particles are far apart, their joint state cannot be described separately since they are correlated in a very strong quantum way. When Alice wishes to send this qubit to Bob, she makes it interact with her share of $|\chi\rangle$ and performs a quantum measurement to obtain two bits of information, which she then sends to Bob by classical means (step 3). These two classical bits tell Bob which of four operations to apply to his share of the state which was formerly $|\chi\rangle$. Once he applies the operation his qubit will be in the state $|\Psi\rangle$ (step 4), while Alice now has an entangled pair of qubits in a generic state very similar to the original $|\chi\rangle$ (the difference is described by the classical bits she sends Bob) and independent of the qubit she teleported to Bob.

per cent correlation, as between the two qubits of $|\chi\rangle$) and that no one else has any information about it (0 per cent correlation). A shared secret key can be used in an array of private key cipher systems including the well-known US Data Encryption Standard (DES) and the only provably secure cipher, the so-called 'Vernam cipher' or 'one-time pad'.

Any information that is stored classically and exchanged between Alice and Bob can in principle be copied by an eavesdropper, Eve, giving her a copy of the information (100 per cent correlation) without a trace. Quantum information, however, cannot be copied or studied without a trace! Several protocols that allow Alice and Bob to produce a common key by the exchange of quantum information or the sharing of entangled particles have been devised. The advantage of distributing the key quantumly is that any eavesdropping or tampering will, by virtue of the uncertainty principle, affect the information being exchanged and the correlation between the keys Alice and Bob share must be reduced. If Eve learns any significant amount of information about the key Alice and Bob share, then the uncertainty principle requires that, with high probability, the keys will not be identical. Alice and Bob can detect this effect of eavesdropping and bound the amount of information Eve has about their key. If the amount of information is small, there are methods, described in the next section, that allow Alice and Bob to distill their keys so that the remaining keys are almost certainly equal and, furthermore, there now is only a negligible correlation to any of Eve's information. If Eve has obtained too much information, they simply abandon the not-so-secret key. These sorts of key distributions have been implemented over distances from 30 cm to tens of kilometres at several laboratories around the world.

Quantum teleportation and key distribution deal with Alice sending information that she explicitly possesses to Bob and vice versa. Suppose, on the other hand, that Alice and Bob need to communicate with each other to figure out some valuable piece of information. For example, Alice and Bob work in different places and their babysitter has just asked Bob whether she would be needed on any of the next seven days since she wishes to go on holiday. Alice and Bob's task is to decide whether there is some day next week when neither will be at home, but each has no prior knowledge of the other's schedule. Neither Alice nor Bob individually possesses this information, but together they possess enough information to figure it out. Thus they must communicate some information to each other to decide the answer. The answer could be a very small amount of information. The *communication complexity* of this distributed computation problem is the amount of information Alice and Bob must send back and forth in order to figure out the answer, which is usually much more than the length of the answer. The communication complexity is the amount of long-distance telephone charges they must accrue in order to solve the communication problem. An easy solution is for Alice to fax Bob her entire schedule and then Bob figures out whether they need the babysitter. In general, if Alice and Bob want to decide whether they need a babysitter in the next N days, using this simple protocol uses of the order of N bits of communication between the two. This seems like a lot of communication in order to compute just one bit of information (YES or NO), but, if Alice and Bob only exchange information classically, it is necessary in some worst-case scenarios. What happens if Alice and Bob can send *quantum* bits of information back and forth? Alice and Bob can in fact solve this problem by exchanging only roughly $\sqrt{N}\log(N)$ quantum bits of information, as shown by H. Buhrman, R. Cleve and A. Wigderson in 1998.

The use of quantum information thus opens many doors for Alice and Bob. For certain problems to do with communication or distributed computation, such as the scheduling problem described above, or the communication between different processors in a multi-processor computer, the amount of information that must be communicated back and forth can be greatly reduced by using quantum bits instead of classical ones. Sending such quantum bits can be done most safely by quantum teleportation. By using quantum information, their communications can also remain private without relying on assumptions about the computational intractability of any problems.

7.4 'In principle' versus 'in practice'

We have already emphasised that the significance of quantum computers (and quantum information physics in general) is not just that it offers a faster way to do some computing, but rather that it offers a qualitatively different way of conceiving of information and computing. Nevertheless, we should beware of a difficulty that often arises in the context of computers, namely that some seemingly promising idea turns out to be completely useless because the difficulties of realising it in a system of useful size were vastly underestimated. Furthermore, quantum coherence is notoriously fragile, especially when complicated quantum systems are involved. Experimenters have to go to great lengths to preserve the coherence of even small quantum systems (such as a few atoms or photons) whose complexity is no greater than a few qubits' worth. So, when we contemplate the quantum computer, these considerations make us smell a rat. Is the striking computational power of the quantum computer actually illusory, since it is based on assuming a degree of precision in the construction of the computer that, for all practical purposes, is impossible to achieve? The suspicion that this was the case certainly kept the interest and excitement of the field somewhat damped down in the early 1990s and, although we are about to discuss the tremendous progress that has been made towards understanding and tackling this issue, it remains the chief reason for caution regarding the future.

Let us examine the requirements on experimental precision for quantum cryptography and quantum computation. Cryptography protocols have to be carefully designed in order to allow a certain amount of experimental imperfection and noise, but it turns out that this is not too severe a problem. When one is setting up a cryptographic key, the main effect of noise is that Alice's and Bob's data are going to differ to some extent, in a roughly random way, irrespective of whether an eavesdropper is present. So how can they distinguish an eavesdropper from random noise? They can't, but as long as the noise level is low enough, they don't need to. If the noise (error probability) is below some small level per bit, then they can rule out strong eavesdropping at least. They then go on to clean up their data by performing 'parity checks' (see below). When they thus correct errors they also shut out any remaining weak eavesdropping because the (limited) amount of data gathered by Eve is effectively dropped by Alice and Bob.

The parity check is a very simple yet fundamentally significant concept of information science. The parity of a string of bits simply indicates whether the string contains an even or odd number of 1s. For example, 01100100 has odd parity (parity equal to 1), whereas 01100110 has even parity (parity equal to 0). In the case of cryptography, Alice and Bob have classical bit strings (the results of their measurements), from which they select agreed random subsets of bits. They each calculate the parity of the subset and publicly report their results. If they found the same parity, then that part of Alice and Bob's data probably contains no errors. It might contain two errors or any even number of errors, which they can test by further random checking. It turns out that such procedures are successful for error rates in the apparatus at the level of a few per cent error probability per bit, which means that a real working system not only could be built but actually has been built with current technology.

The working cryptography systems are based on using photons as qubits. Weak light pulses containing single photons are sent through a specially designed interferometer in Alice's laboratory, and then transmitted several kilometres down standard fibre-optic telecommunications systems to Bob, who has a similar interferometer. Taking advantage of various ingenious methods to enhance the preservation of the photons' quantum states, the tolerated bit error rate of a few per cent can be achieved.

This is in contrast to the situation with quantum computers, because no such computers exist and they will not for some time. However, current technology does permit us to build few-qubit systems that allow the basic concepts of quantum information processing to be demonstrated and from which we can learn how to go further.

In principle there are myriad ways in which one might conceive of an experimental system, but in practice only a few satisfy the severe requirements of sufficient complexity and sufficient controllability. Currently there are two systems in which three or so qubits can be manipulated, these are the ion trap and the nuclear magnetic resonance (NMR) spectrometer. The former takes advantage of high-precision atomic-physics techniques such as laser cooling to allow one to manipulate the motion and internal states of individual charged atoms (ions). The ions are confined by a set of electrodes in high vacuum and addressed by laser beams focused onto them. The latter uses a standard NMR spectrometer and manipulates the nuclear spins in a simple molecule using pulsed magnetic fields. These two approaches are to some extent complementary, in that they have different strengths and weaknesses.

The former allows complete interrogation of a single line of qubits, whereas the latter uses a liquid sample containing billions of molecules and interrogates the average state. To date, thorough manipulation of two qubits is more or less routine for NMR work and limited manipulation of three to seven qubits has been reported. Meanwhile, a single ion-trap experiment has achieved manipulation of two to four ions, though not yet a completely general set of operations. Typically the precision of these experiments allows around a hundred quantum logic gates such as exclusive-OR to be applied before the state of the system is lost to noise and imprecision.

Needless to say, all this is a very long way from the level of quantum computing we would need in order to achieve a real rival to classical methods. In order to see how far, we can consider an example task for a quantum computer. We will take this to be the factorisation of a thousand-digit number. Although we do not expect factorisation to be the main use of quantum computers in the future, this is a quantum algorithm we understand and it might give us a feel for the size of machine we need to envisage in order to do useful processing which could not be done on classical computers. Shor's algorithm for this task would require the level of noise in the operations, and of coupling between the qubits and their environment, to be kept below approximately 10^{-17}. This level of precision is so difficult to attain that one can rule it out as more or less impossible. The reason has to do with the connection between legitimate operations and troublesome ones: in order to manipulate qubits, there must exist a physical interaction between them and the controlling machinery, but that implies that there is also a coupling between the qubits and other stuff, such as electrical noise or, when all else has not failed, quantum-mechanical fluctuations of the vacuum.

The situation looks at this stage like a house of cards: we can build a layer or two, but when one contemplates building a really tall house, the task seems hopeless. Nevertheless, the estimate we just made could equally well have been applied to classical computers. They routinely complete computations requiring this number of bits and gates, without making an error (well, most of the time, anyway!). How is their remarkable reliability possible? The essential ingredient is that every bit in a classical computer is under scrutiny all the time! The on-chip transistor circuits play a role equivalent to that of the spring in a mechanical switch, forcing the switch, or in this case the bit value, one way or the other. Any small departure is 'detected' and forcefully suppressed by such strong 'springs'. Unfortunately, no such scrutiny is permitted in a quantum computer! To examine a qubit is to render it useless for quantum computation. What we need is a much more subtle approach, whereby we do not let our right hand know what our left hand is doing: we would like to detect erroneous changes in the quantum computer's state without learning anything about the state itself. We now know how to do this.

Quantum error correction (QEC), depicted in Figure 7.4, is a set of

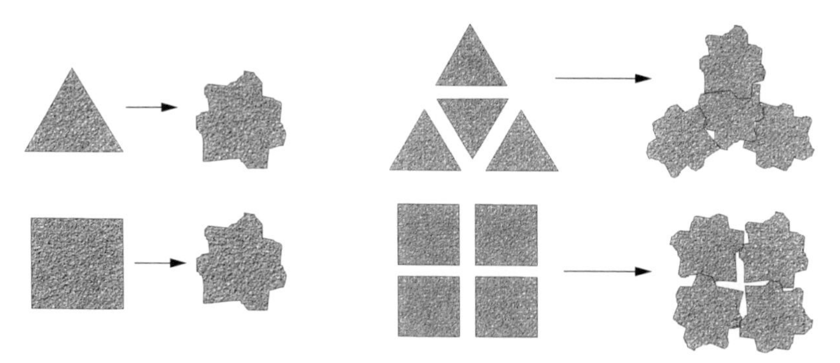

Figure 7.4. Quantum error correction. A single bit of quantum information is represented by two distinguishable states, here symbolised by a triangle and a square. Noise will blur these states, making them indistinguishable. However, if we use four qubits together, we can construct joint states in which the quantum information is stored communally. Even though every qubit becomes blurred, the stored joint state is still detectable in the pattern among the blurred qubits (in practice, the pattern is parity information).

powerful and elegant ideas that spring rather naturally from the union of information science with quantum theory. A central ingredient is the operation of parity checking, which can be adapted to the quantum context. This is a new 'quantum' version of parity that has no analogue in classical computation. In this approach we need to use quite subtle and, as it turns out, highly entangled, quantum states, but fortunately their exact construction can be based on the classical theory of error correction which has been studied for fifty years. The extraordinary thing about all this is that the carefully constructed entangled states (called quantum error-correcting codes) do not require exponentially expanding resources in order to achieve exponentially suppressed noise levels.

At this point the reader should still not feel altogether happy about building the house of cards. Although we have introduced corrective measures, what if they themselves are faulty, as they must be in any real system? Even this consideration can be met. We need to invoke a further set of new and subtle ideas, which together go by the name of 'fault tolerance'. The basic problems we face are that a 'correction' based on bad parity information will actually make the situation worse, not better, and even a perfect gate operation will couple errors that occurred previously from one qubit to another, thus spreading the noise. The essence of the answer is two-fold: first, the parity information we require for QEC is essentially classical once we have obtained it, so it can be extracted repeatedly until a consistent result is obtained and only then is the computer corrected. Secondly, because QEC is exponentially efficient, it can mop up not only the noise in the computation itself, but also that generated by the checking operations, as long as they do not amplify noise exponentially. Such an amplification is avoided by careful construction of the quantum networks, restricting routes for the propagation of errors.

With all these ideas working together, the 'realistic' quantum computer looks very different from the idealised noise-free one. The latter is a silent shadowy beast at which we must never look until it has finished its computations, whereas the former is a bulky thing at which we 'stare' all the time, via our error-detecting devices, yet in such a way as to leave unshackled the shadowy logical machine lurking within it. For every elementary logic gate of the logical computation, the corrective procedures involve thousands of gates and thus dominate the machine, but we have gained a great deal because the machine can now tolerate noise in every one of these gates at the level of 10^{-5}. This degree of precision is achievable, in contrast to the figure of 10^{-17} which we previously had to contemplate.

7.5 Future prospects

When the physics of computation was first investigated systematically in the 1970s, the main fear was that quantum-mechanical effects might place fundamental bounds on the accuracy with which physical objects could realise the properties of bits, logic gates, the composition of operations and so on, which appear in the abstract and mathematically sophisticated theory of computation. Those fears have been proved groundless. As we have explained, quantum mechanics, far from placing limits on what classical computations can be performed in nature, permits them all and, in addition, provides entirely new modes of computation, including algorithms that perform tasks that no classical computer can perform at all (secure key distribution) or that a classical computer can perform, albeit not efficiently (factorisation). Experimental and theoretical research in quantum information processing is accelerating world-wide. New technologies for realising quantum computers are being proposed and new types of quantum information processing with various advantages over their classical versions are continually being discovered and analysed.

The current challenge is not to build a full quantum computer right away but rather to move from the experiments in which we merely observe quantum phenomena to experiments in which we can *control* these phenomena. This is a first step towards quantum logic gates and simple quantum networks. The next challenge is to scale quantum devices up. The more components the more likely it is that quantum computation will spread outside the computational unit and will irreversibly dissipate useful information to the environment. Thus the race is to engineer sub-microscopic systems in which qubits interact only with themselves, not with the environment. New techniques, such as quantum error correction and fault-tolerant computation, together with new technologies will allow us to achieve this task in the not-too-distant future. At this point new devices, such as ultra-precise quantum clocks and entanglement-enhanced frequency standards, will supersede the existing ones. New quantum sources of light, better and cheaper photo-detectors and quantum repeaters will make quantum cryptography a serious alternative to more traditional ways of encryption. Re-transmission via satellite can even make quantum cryptography suitable for long-distance communication. The next millennium will witness computer technology departing from silicon and new quantum algorithms run on quantum processors. Quantum computers will eventually become a reality with a number of useful applications. There is no way we can predict them now. Imagine Charles Babbage being asked about the future of his analytical engine – would he have been able to predict word processors, computer games and the Internet? Whatever these applications might be, one of them will be an efficient simulation of complicated quantum phenomena. This will help us to set up new experiments that will refute quantum theory and will let us learn more about the laws of physics. These breakthrough discoveries and their implications for computations will be nicely summarised by our descendants in the millennium edition of the *Philosophical Transactions* in the year 3000.

7.6 Further reading

More information is available at http://www.qubit.org and in the following.

Bennett, C. H., Brassard, G., and Ekert, A. K. 1992 *Scientific American* October, 50.

Ekert, A. and Jozsa, R. 1996 *Rev. Mod. Phys.* **68**, 733.

Mosca, M., Jozsa R., Steane A. and Ekert, A. 2000 *Phil. Trans. Roy. Soc., Special Millenium Issue.*
1998 *Physics World* **11** March (Special issue on quantum information).
1997 *SIAM Journal on Computing* **26** (special section on quantum computation), 1409.
Steane, A. 1998 *Rep. Prog. Phys.* **61**, 117.

8
Magnets, microchips and memories

Russell P. Cowburn

Nanoscale Science Group, Department of Engineering, University of Cambridge, Trumpington Street, Cambridge CB2 1PZ, UK

8.1 Introduction

The latter half of the twentieth century has been marked by a ten-million-fold increase in the engineer's ability to store information magnetically (e.g. on computer hard-disk drives). Without this, the computing revolution which the developed world has experienced would not have happened. The exciting and challenging question which faces us now is that of how to keep the growth going. The emerging fields of nanotechnology and quantum engineering may provide solutions that will satisfy engineer and consumer alike and change the very way in which society understands itself.

Magnetism has been known to mankind for many hundreds of years and has from the beginning been essential to navigation. The first-century-BC scholar Lucretius wrote of the magnetic properties of lodestone, although his understanding of magnetism was perhaps better informed than that of the thirteenth-century-AD scholar Bartholomew the Englishman, who assures us that 'This kind of stone restores husbands to wives and increases elegance and charm in speech. Moreover, along with honey, it cures dropsy, spleen, fox mange, and burn.'!

The London physician William Gilbert was the first to make serious inroads into an understanding of magnetism. Gilbert, who was a Fellow of St John's College, Cambridge, Physician to Elizabeth I and the founder of a precursor to the Royal Society, published his great work *De Magnete*

Magneticisque Corporibus et de Magno Magnete Tellure Physiologia Nova (*On the Magnet: Magnetic Bodies Also, and On the Great Magnet the Earth; a New Physiology*) in 1600. Gilbert's eminent contemporary Galileo Galilei read the work and consequently described Gilbert as being 'great to a degree that is enviable'. The portrait shown in Figure 8.1(a) was left to Oxford University after his death and bears the inscription 'Gilbert, the first investigator of the powers of the magnet'.

Gilbert lived too soon, however, to see the real story of magnetism unveiled, which had to wait another 200 years for the arrival of Maxwell and Faraday. Their great contribution was to unite electricity and magnetism into 'electromagnetism'. Another revolutionary application thus appeared from magnetism – the dynamo, or electrical generator, and its sister machine, the electrical motor. It is from these that the twentieth century remodelling of society by electricity and electronics flowed.

In addition to navigation and electromagnetism, magnetism has had an impact on human history in a third way. This third revolution was started very quietly at the end of the nineteenth century in the laboratories of the Copenhagen Telephone Company by Valdemar Poulsen. Poulsen reasoned that people would find a device that could record telephone messages useful and so invented the 'telegraphone' (Figure 8.1(b)), a precursor to the tape recorder but using piano wire instead of magnetic tape. The Austrian Emperor Franz Joseph saw the telegraphone at the Paris Exposition in 1900 (where, incidentally, it won the Grand Prix) and recorded a message. That message still exists today; it is the world's oldest magnetic recording.

The irony is that the telephone-recording application envisaged by Poulsen failed to materialise in most people's homes until the 1990s, some 100 years later. Nevertheless, the principle that magnetism could be used in a *memory* device had been demonstrated. This principle is now used in every aspect of modern life, be it storing information on your computer, storing bank records, recording hospital scans, preparing television and radio programmes, preparing music CDs or saving personal details on the backs of credit and debit cards.

8.2 Magnetic memory

Scientists express the key concept that magnetic materials possess a memory function by invoking a *hysteresis loop*. Figure 8.2(a) shows a

(*a*)

(*b*)

Figuere 8.1. Views from antiquity. (a) William Gilbert, the father of magnetic research; and (b) Poulsen's telegraphone, the first magnetic recording machine (from *Scientific American*, 22 September 1900).

Figuere 8.3. A survey of modern data-storage technology. From the left, an old 5.25″ floppy disk (100 kbytes), a modern 3.5″ disk (1.4 Mbytes), a Zip disk (100 mbytes) and a magneto-optical disk (up to 640 Mbytes). In the centre foreground is an open hard-disk drive (up to 27 Gbytes).

ungrammatically called the 'media' and consists of a rigid, rapidly spinning plate on which a thin film of a cobalt–chromium alloy has been placed. In the foreground is an arm with a small sensor at the end, called the 'head'. A combination of the disk spinning and the arm rotating allows the head to reach any part of the disk's surface. The head is comprised of two parts: a write head, used to magnetise the disk locally either positively or negatively, and a read head, used to sense the magnetic field coming from the disk and hence determine the direction (fully positive or fully negative) in which a region has previously been magnetised. These two parts respectively achieve the data recording and play-back functions.

A stream of 1s and 0s is thus written onto the disk as a circular track of positive and negative magnetisations in a row. Figure 8.4 shows an actual picture taken by a special microscope that can 'see' positive and negative magnetisation. Each magnetised region corresponds to one data bit, so the storage density of the disk is determined by how large each of the written bits is. The phenomenal growth in storage density described previously has until now been achieved simply by repeatedly reducing the dimensions of the written bit. However, fundamental physical limitations that are expected to prevent the bit size shrinking much further now stop engineers from sleeping as soundly as they used to.

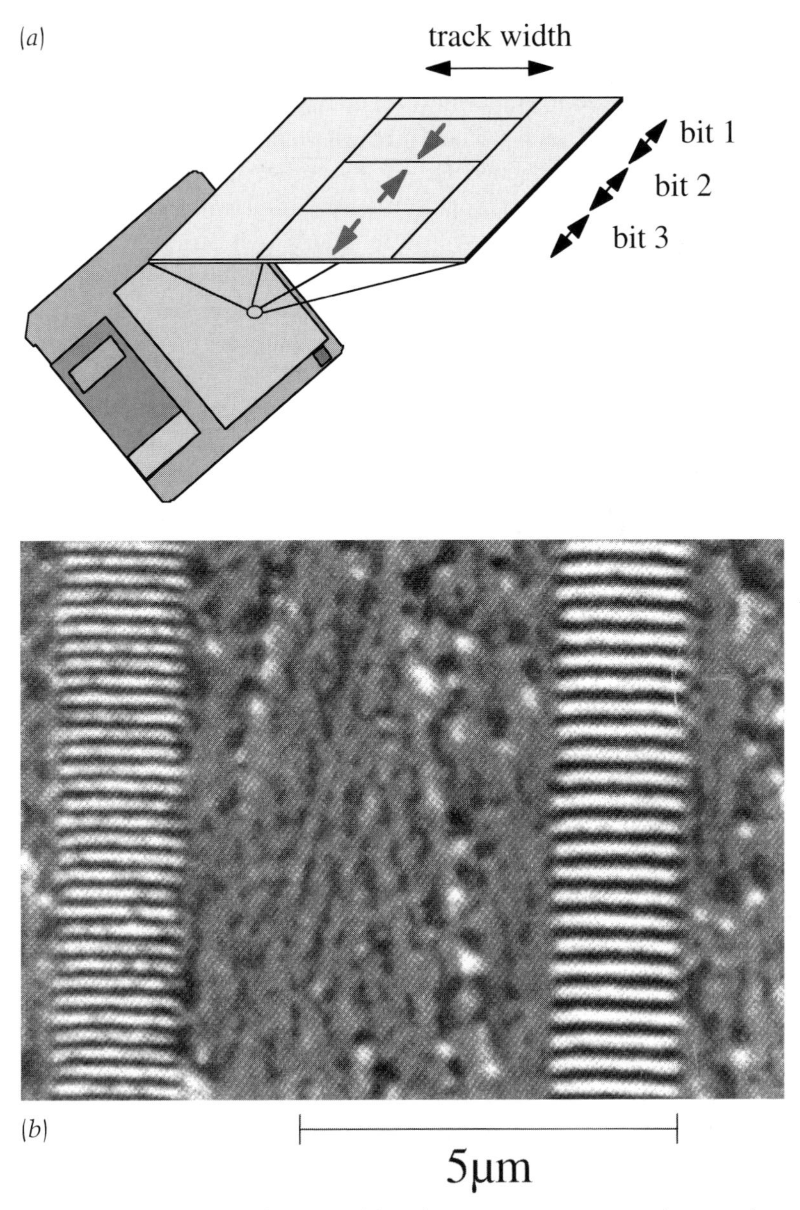

Figuere 8.4. Magnetic-disk storage: (a) a schematic representation of written bits and (b) a magnetic microscope's view of two tracks of written bits. The black bands represent 1s and the white bands represent 0s (from Malhotra *et al. IEEE Trans. Magn.* **33**, 2993 (1997), © 1997 IEEE).

The first of these limitations is due to the material from which the media is made and is known as the 'media-noise problem'. The hard-disk media is not made from a uniform sheet of magnetic material. It is rather comprised of a large number of small magnetic islands (or grains) tightly packed together in a non-magnetic sea. The small non-magnetic regions between the grains prevent one bit from 'leaking' and growing in size. If the bit size is large relative to the grain size, then a well-defined rectangular bit is possible. If, however, the bit size is comparable to the grain size, then the bit has rough, ill-defined edges, which increases the noise in the sensor. The number of erroneously read bits therefore increases.

One approach to the problem of media noise is to find new materials for the media that have smaller grains. This, however, leads on to a second problem, which is called 'superparamagnetism' and comes about because of temperature. Just as a molecule at a finite temperature vibrates and rotates, so the direction of magnetisation in a magnetic material must fluctuate. Now, in a large bar magnet this does not matter because the size of the fluctuation is tiny. The smaller the magnet becomes, however, the stronger the fluctuations become. Current hard-disk-media grains are still large enough that the fluctuations in magnetisation can be forgotten. If, however, the grain size is reduced much further, then the fluctuations in magnetisation will become so large that they will sometimes be able to turn the magnetisation through 180°, converting a '1' into a '0' and corrupting the data.

A third problem is concerned with reading the data back from the disk. The smaller the bit size, the smaller the magnetic field which comes out of it. An increasingly sensitive detector is therefore needed for the read head and the read head must 'fly' increasingly close to the surface of the disk. Flying the read head in even today's hard disks is like flying a Boeing 747 airliner at 500 mph at a distance of less than an inch above the ground, counting the blades of grass as it goes!

8.3 Magnetic nanotechnology

Nanotechnology is a field with immense future potential that has arisen largely during the 1990s. It is the art and science of manipulating and using material on the nanometre scale (1 nanometre, or 1 nm, is one thousand millionth of a metre, or about five atoms in size). There are two main workhorses in nanotechnology: the electron microscope and the scanned probe.

Figuere 8.5. Excerpts from nanotechnology. (a) Part of Gilbert's *De Magnete* written in our laboratory by electron-beam lithography, using penstrokes 30 nm wide. The letters are small enough to copy all twenty volumes of the *Oxford English Dictionary* onto the space occupied by a single capital letter on this page. (b) A scanned probe cantilever with a close-up of the sharp tip in the inset (picture by J. Barnes). The extreme sharpness of the tip allows objects as small as a single atom to be seen and manipulated.

The electron microscope is used to perform *electron-beam lithography*, which is essentially a photographic process. Instead of producing glossy pictures, however, it produces physical structures by using the sub-nanometre-size focused electron beam to draw out the shape of the desired nanostructure onto a photographic layer. Several chemical steps, directly analogous to developing a camera film, are then used to convert the exposed shapes into real physical structures, which, for the current state of the art, can be as small as 5 nm. Figure 8.5(a) shows an example of electron-beam lithography, whereby a passage from Gilbert's *De Magnete* has been written onto a piece of silicon, using 30-nm-wide penstrokes. The letters are small enough to copy all twenty volumes of the *Oxford English Dictionary* into the space occupied by a single capital letter on this page!

The scanned probe is a development of the scanning tunnelling microscope which won Binnig and Rohrer half of the Nobel Prize for Physics in

1986 for work that they had carried out in the IBM Zürich Research Laboratory. In scanning tunnelling microscopy, and its sister technique of atomic-force microscopy, an ultra sharp point (see Figure 8.5(b)) is used to 'feel' the shape of an object as it is scanned over a surface. These microscopes can see single atoms and so provide an excellent way of looking at nanostructures. Moreover, the sharp point can be used to pick up and move pieces of material as small as a single atom, so they can also be used for nanoscale building.

The technology of nanometre-scale construction is still not fully mature. One of the greatest challenges facing the field is that of the speed of construction. The text shown in Figure 8.5(a) took approximately 10 s to write. Although this may seem reasonably fast, some of the best commercial applications would need production to be one hundred million times faster than this. Ways of building nanostructures more rapidly will therefore need to be vigorously researched this decade if these devices are ever to leave the laboratory.

The reason for discussing nanotechnology in this article is because of a very important theoretical prediction made by William Fuller Brown Jnr in 1968. Brown was one of the great pioneers of the field of *micromagnetics*, the mathematics of magnetism on a microscopic scale. In order to understand his prediction, we need to understand the various forces inside a magnet. In a typical magnet of any size there are usually at least two different internal processes competing with each other. These are called *demagnetisation* and *exchange* and are illustrated in Figure 8.6. In the case of demagnetisation, the magnetisation at the edges of a magnet 'sticks out' through the sides, leading to the appearance of magnetic North and South poles on the sides, just as in a conventional bar magnet. These poles are sources of magnetic field lines, which as well as passing *outside* the magnet (as seen in Faraday's famous iron-filing experiment) also pass backwards through the *inside* of the magnet. This field, called the *demagnetising field*, opposes the direction of magnetisation and tries to rotate the magnetisation in such a way as to reduce the strength of the poles. One way of doing this is by forming the whirlpool (or 'vortex') pattern shown in Figure 8.6(a). Such a pattern does not have any poles on the edges because the magnetisation is always parallel to its nearest edge, so the unfavourable demagnetising field is not generated. If demagnetisation were the only process at work in a magnet, then permanent magnets and magnetic memory would not exist, because the demagnetising field could always cause the magnetisation to collapse.

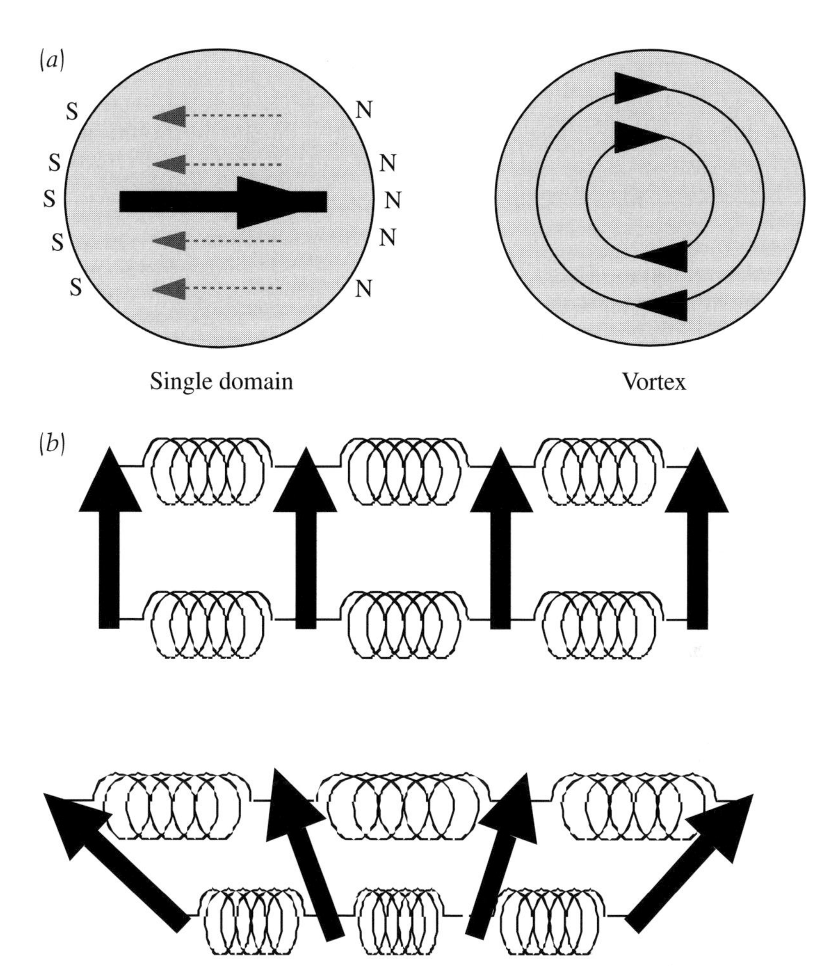

Figure 8.6. The two opposing forces in nanomagnets: (a) demagnetisation, whereby the surface poles (N and S) create an internal demagnetising field (dashed line), which causes the magnetisation (solid line) to break into a vortex; and (b) quantum mechanical exchange, which appears as springs between spins, keeping neighbouring spins parallel. Magnetism would be a much less interesting (and certainly less useful) phenomenon without this quantum-mechanical effect.

Fortunately, demagnetisation is opposed by the second internal process, exchange. This is a quantum-mechanical effect and is ultimately responsible for magnetism. Quantum mechanics is one of the great success stories of twentieth-century physics, achieving popular fame via Schrödinger's Cat, and is a highly counter-intuitive, but accurate, description of very small objects such as electrons and atoms. What we have so far referred to as 'magnetisation' actually comes from a quantum property of electrons called 'spin'. Each electron can be thought of as a tiny bar magnet pointing either up or down, depending on the spin. One of the basic quantum laws says that no two objects can have the same quantum description as each other. This means that electrons of the same spin direction try to stay away from each other, which is actually a very good thing because when two electrons do approach each other there is an electrostatic-energy penalty to pay ('like charges repel'). If, however, the electrons have their spins pointing in different directions, then quantum mechanics does not try to keep them apart because there is already some difference in their quantum descriptions. Consequently, the electrons encounter each other frequently and feel the full strength of the electrostatic repulsion that exists between two negative charges. Overall, the spins in a magnetic material attempt to stay aligned parallel with each other whenever possible in order to minimise energy. If one tries to push neighbouring spins to point in different directions, then it is as if there were springs between them, as shown in Figure 8.6(b). The springiness of this exchange, or exchange stiffness as it is correctly known, opposes any action that prevents the spins being parallel to each other.

Brown understood the perpetual competition between demagnetisation and exchange and it led him to what has become known as Brown's Fundamental Theorem. Brown realised that, in large magnets, demagnetisation will prove to be the stronger competitor because of the large surface area of the poles. Conversely, as a magnet is reduced in size, there should come a point at which exchange will gain the upper hand. Very small magnets cannot therefore ever be demagnetised and hence must adopt the so-called single-domain state. The importance of this point for magnetic data storage cannot be understated. Nanometre-scale magnets (for this is the lengthscale on which quantum mechanics becomes the dominant force) are the ideal data-storage device, for they do not lose their memory. Only in the last few years has nanotechnology reached a sufficient state of maturity to allow researchers to begin making and testing these tiny magnets. These are first steps into the field of *quantum engineering*, in which nanotechnology is used to make devices small enough that they allow access to the quantum world.

The application of nanotechnology to magnetism has so far been presented in the context of one very specific application, namely making new data-storage media. The issue is, however, much wider than this. The latter half of the twentieth century has seen the invention of hundreds of different magnetic materials. Given that there are only nine or so commonly used magnetic elements, nearly all of these materials have been made by alloying the nine magnetic elements with each other and with non-magnetic elements. Nanotechnology promises a new generation of artificial magnetic materials. Each nanomagnet is analogous to a giant artificial atom, so one is now free to build new materials, giant atom by giant atom. Magnetic nanotechnology will allow designers to specify the precise magnetic properties they require for future magnetic-memory technology and receive a sample of a new material made of artificial giant atoms that possesses just the desired properties.

The quantised magnetic disk is one example of a new hard-disk media created by nanotechnology. Figure 8.7(a) shows a picture of such a disk. The quantised magnetic disk, which is currently available only in research laboratories, is made of millions of artificially created magnetic islands or pillars. These identical islands do not grow by chance; rather, each is individually placed by the material designer. It is expected that quantised magnetic disks, or patterned media as they are also known, will be used in future generations of hard-disk drives.

Nanotechnology will almost certainly have an impact on the read/write heads in hard-disk drives. A more immediate improvement that can be made is described in the next section, but ultimately the single-atom resolution of the scanned probe could be used for reading and writing with ultimate precision. One possible configuration for this is shown in Figure 8.7(b), in which a pointed nanomagnet is mounted on the end of a scanned probe cantilever. If the tip is lowered towards the surface of the hard disk (which, in this case, is assumed to be a quantised magnetic disk) then the magnetic fields coming from the tip will write a data bit onto the disk. If the tip is withdrawn slightly, then the attraction between the tip and the written bit bends the cantilever in a way that can be used to read the bit. The picture of conventionally written hard-disk bits shown in Figure 8.4(b) was obtained in just this way.

8.4 Spintronics

Spintronics, or magnetoelectronics as it is also known, came about thanks to a very important discovery made in 1988 by the research teams of Albert Fert in Paris and Peter Grünberg in Julich. Fert and Grünberg asked a

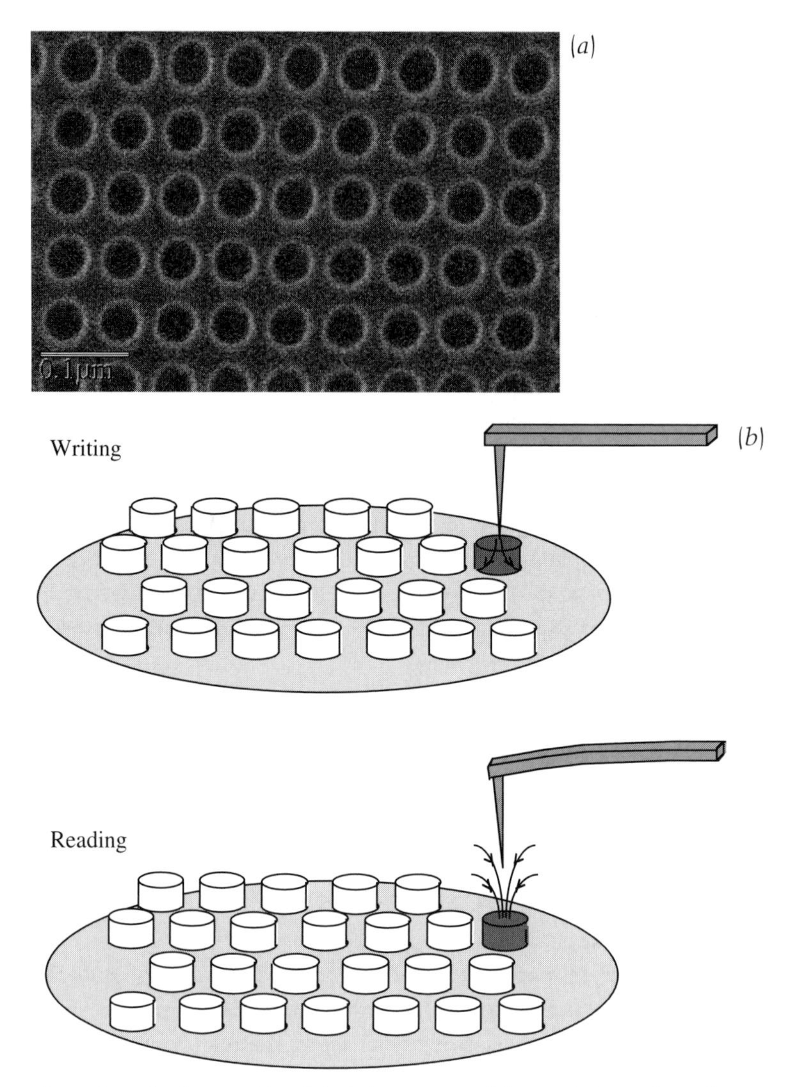

Figure 8.7. The future of hard-disk data storage. (a) Part of the surface of a quantised magnetic disk. Each pillar is 40 nm in diameter and could store one bit of data. A typical hard disk could contain 300 thousand million identical pillars. (b) Writing and reading a quantised magnetic disk with a magnetic scanned probe. When the sharp probe is close to the magnetic pillars, it can write data onto them. When it is further away, the magnetic field from the pillars attracts or repels the probe, allowing data to be read back. The sharpness of the tip allows information to be very tightly packed.

simple question: 'how does the electrical resistance of two individual magnetic films, placed on top of each other, change in a magnetic field?'. To their surprise, they found that, instead of the tiny variation in resistance which usually arises when a magnetic field is applied to a single magnetic layer, they obtained an *enormous* change. The new effect was graphically named giant magnetoresistance. A vitally important principle had been demonstrated: electronic circuits and magnetic materials are very natural relations. Hence the field of spintronics (magnetic spin plus electronics) was born.

Figure 8.8 shows a typical way of observing giant magnetoresistance, by passing a current from one piece of magnetic material into another, via an intermediary non-magnetic layer. If the two pieces of material are magnetised in the same direction (Figure 8.8(a)) then good conduction occurs. If, however, the pieces of material are magnetised in opposite directions (Figure 8.8(b)) then very little current flows, and a high electrical resistance develops. The device which results is called a *spin valve*, because it is like a water valve. The directions of magnetisation act as taps for the flow of electrical current. The spin valve can serve as a very sensitive magnetic-field sensor and also as a memory cell. Further technical details of how this remarkable effect works can be found in the Millennium Issue of *Phil. Trans. Roy. Soc.* A.

The consequences of the effect of giant magnetoresistance lie both in the present and in the future. Since 1999, most of the world's new hard-disk drives have contained giant magnetoresistance read heads. Using a giant magnetoresistance spin valve as a field sensor allows a sufficiently strong electrical signal to be obtained from correspondingly weaker magnetic fields and hence from much smaller bits.

The future impact of giant magnetoresistance will come from new generations of magnetic memory devices. The most promising of these is the magnetic random-access memory (MRAM) chip, currently being developed by companies such as Honeywell and IBM. With MRAM one seeks to replace electronic memory chips found in computers by magnetic chips. Each bit would be stored in a nanometre-scale piece of magnetic material and then giant magnetoresistance, or a related effect, used to read the data back in electronic form. MRAM offers many advantages over conventional memory. First of all, it could achieve very high storage densities, because only one small magnetic element is required per data bit. Secondly, MRAM retains its memory when a system is powered down, so computers would

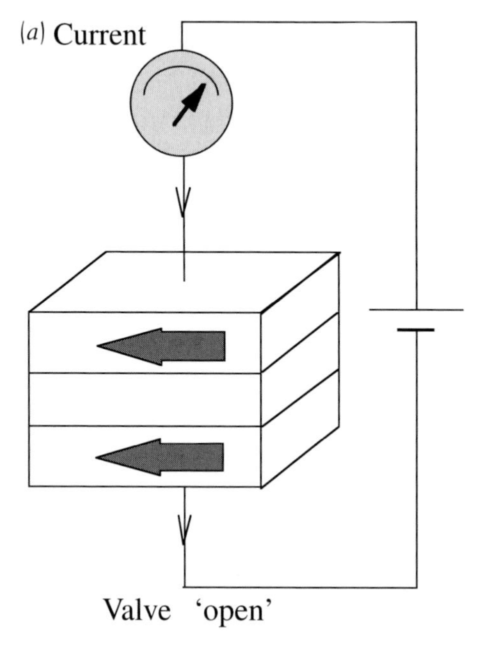

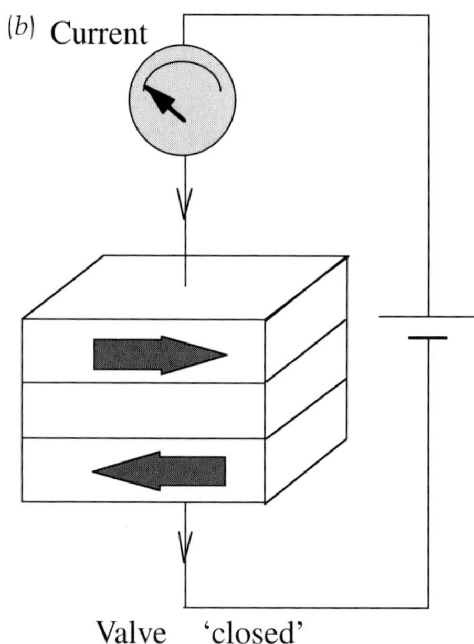

Figure 8.8. The giant magnetoresistance spin valve: (a) parallel alignment of the two magnetic layers, leading to a low resistance; and (b) anti-parallel alignment, leading to a high resistance. It is called a *spin valve* because it is like a water valve: the directions of magnetisation act as taps for the flow of electrical current. The spin valve can serve as a very sensitive magnetic-field sensor and also as a memory cell.

not need lengthy hard-disk re-boots every time they are switched on. The rapidly growing portable computer market desperately needs this type of memory which can store information without consuming any power. MRAM is potentially very fast, operating in principle on a sub-nanosecond timescale. A final advantage is that it is not susceptible to radiation damage in the way that semiconductor memory is, so it would be excellent for military and space applications.

MRAM is now established as a future technology and so it is most probable that twenty-first-century computers will use magnetic memory. Less certain at this time is how much further the conversion of electronics to spintronics will go. Just as hard-disk data storage density has been expanding rapidly, so has microprocessor power and memory-chip capacity. The magnetic transistor has already been demonstrated and our laboratory is currently working on magnetic logic gates. A wholesale replacement of electronics by spintronics might be one way in the future to allow components to continue shrinking.

8.5 The way ahead

The hard-disk data-storage industry currently shows no signs of slowing down. Current media will be pushed for a few more years, so we should soon be purchasing hard-disk drives with capacities around 60 GBytes (today's limit is 27 Gbytes) that work in essentially the same way as today's drives. After this, today's continuous media will be replaced by patterned media. I believe the ultimate storage density of this technology to be around 400 Gbits per square inch (corresponding to 20-nm-sized bits), which compares with today's density of 10 Gbits per square inch. A typical desktop hard drive would in this case be able to store around 1000 Gbytes. Two extrapolation lines have been added to Figure 8.2(b) showing that we could expect this limit to be reached sometime between 2008 and 2017.

What happens after that? The real problem with current magnetic storage is that it is intrinsically two-dimensional. Data are stored only on the surface of a disk, which is very wasteful. Three-dimensional data storage has to be the future for the third millennium. There is some chance that it will still use a magnetic principle, but that is not yet certain. The advantages of going to three dimensions can be demonstrated by a simple calculation. My prediction of 400 Gbits per square inch as the ultimate maximum was made assuming one data bit every 40 nm on the surface of a disk. If, now, the same size of bit is used, but this time inside a three-dimensional solid, then the equivalent storage density is 260 000 000 Gbits per cubic inch! In the ultimate limit of one atom representing one data bit, a piece of material the size of a sugar lump could store 10^{25} bits, or 10 000 000 000 000 000 Gbits. The numbers are unimaginable! The big problem facing this idea is one of addressing: how exactly does one read from and write to the atom in the middle of the block? Although it is a mighty technological challenge, I believe that a solution could be found at some point in the twenty-first century. The answer might well involve using electromagnetic waves, especially if a cheap X-ray laser were to become available.

I firmly believe that magnetic RAM chips will very soon become a reality. The technical difficulties involved are, in my opinion, surmountable. The biggest difficulties are institutional. A semiconductor fabrication plant is an unbelievably expensive purchase. Today's price tag is around £1.3 billion and rising. The inertia arising from this scale of investment creates a very understandable conservatism within the industry, meaning that any progress has to be made by evolution rather than by revolution. Perhaps this will be less of a problem in the future. Should the semiconductor industry itself undergo a major change in structure (perhaps moving to many small manufacturing sites due to advances in fabrication technology and a market requirement to diversify the range of products), then perhaps new ideas such as MRAM will have more latitude to develop.

One of the interesting aspects of the development of the hard-disk industry on the one hand and MRAM on the other is that the two technologies are on converging paths. Conventionally, hard-disk storage is cheaper (per bit) than RAM because the data-storage sites do not need to be predefined. If, in the future, one sees a move towards patterned media and semiconductor RAM chips become magnetic RAM chips, then the only difference between the two will be the addressing method: MRAM will use a mesh of wires, whereas the quantised magnetic disk will use a flying head. The problems facing the flying head will become increasingly acute, however, as the bits become smaller. There could soon come a point at which it is simpler to overlay the quantised magnetic disk with a mesh of wires and address the individual bits that way. At that point, hard-disk and RAM technologies will have merged.

We may, in the future, see new applications of magnetism in addition to data storage. Many science-undergraduate coffee rooms buzz with talk of a quantum computer. The quantum computer, which is currently only a theoretical construct, uses the uncertainty of the quantum world to perform many calculations simultaneously. Unfortunately, the theory is still far ahead of the practice. Because the theory is so general, it is currently not even known which branch of science to use to implement the idea. Magnetism might yet prove to be the most favourable. Very small nanomagnets could perhaps be used, although probably only at very low temperature. The major difficulty facing the quantum computer is that of how to control errors. The problem becomes less acute the smaller one goes in size, so nanotechnology will certainly be needed. One idea currently being considered is to use the very small magnetic moment contained in the *nucleus* of the atom. This is not so far-fetched as it might seem, for already hospital magnetic-resonance-imaging scanners do precisely this. A related but separate development, which is currently in its infancy, uses a scanned probe cantilever such as the one shown in Figure 8.5(b) to perform magnetic resonance imaging across very small objects. The goal of this work is to visualise the individual atoms in a molecule such as a protein by using magnetic resonance imaging. This would be the ultimate biochemical analysis tool.

One issue concerning data-storage technology in the future remains to be addressed and in many ways it is the most important: why do we need so much storage capacity? A modest hard drive can today store all of the *Encyclopaedia Brittanica* with plenty of space left over, but how many PC owners believe they will read even all of that in a lifetime? In my opinion, we need more storage capacity *not* to do more of the same, but to do *different* things. I tentatively suggest that, if storage capacities grow by many orders of magnitude (as is envisaged with three-dimensional storage), then two major changes in society will come about as a result.

The first is that the nature of information will change. Traditionally we think of information as being words. Until recently, computer data storage could only really handle text. Graphics could be stored, but they filled the disk space very rapidly. If the available space were much, much greater, then all kinds of different forms could be used for communicating: pictures, sounds, animation, etc. Just as poetry is better for expressing love than it is for teaching somebody how to use a video recorder, we will be free to choose the form of information most suited to the content. If this is combined with widespread use of e-books (a form of portable computer that might soon replace books) then the nature of information really will change. The currently separate roles of authors, artists, actors and musicians could all merge. This can happen only if we can store it all.

The second change in society which I see coming about as a result of an unlimited ability to store information is perhaps the most important and is a change in the way we understand history. The second millennium has witnessed an enormous breaking down of *spatial* barriers. For most of the inhabitants of the earth during the eleventh century, the neighbouring village was a distant place, seldom visited. Most inhabitants of the earth during the early twenty-first century will have visited most of the major cities in their country, many of the countries in their continent and several countries in the world. In contrast, the *temporal* barriers which divide one generation from another remain and, if anything, are more pronounced than they were a millennium ago. Mass data storage is not a time machine. It is, however, a temporal telescope, allowing us to see clearly into the past. Suppose that every detail of life in the twenty-first century were able to be stored. Would not the inhabitants of the earth during the twenty-second century understand their past (and hence, to some extent, themselves) more clearly? Just as science has been the most successful academic discipline of the latter half of the second millennium, giving us great insights into the natural world and an ability to control it for our benefit, so perhaps will history become the dominant academic discipline of the third millennium. Mass information storage will provide an abundance of primary historical sources, allowing the historian to delve deeply into the workings of the human world, just as the development of the experimental scientific method provided the unbounded number of primary scientific sources which allows science to work so well. The development of the *form* of information will be essential in this. Anyone who has wasted an afternoon drowning in Internet detritus looking for a single piece of information will know that more storage of data does not necessarily make information more accessible. I would hope that future progress in information technology, building on the first attempts of the twentieth century at virtual reality, would allow the historian really to spend a day in the life of a person of (future) antiquity. Advanced bio-interfaces would allow that person's thoughts to be rethought, their feelings refelt and their full world-view re-experienced. To understand history correctly is to understand the influences which have made us the people we are today and therefore to be able to weigh our opinions more freely and reasonably. If future progress in magnetic nanotechnology can allow mankind to do that, then magnetism does indeed have an attractive future.

8.6 Acknowledgements

This work would not have been possible without the assistance of my colleagues Professor Mark Welland, Dr Kunle Adeyeye and Denis Koltsov and the support of St John's College, Cambridge and The Royal Society, London.

Further reading

Cowburn, R. P. 2000 Property variation with shape in magnetic nanoelements. *J. Phys.* D. **33**, R1.

Cowburn, R. P. 2000 The attractions of magnetism for nanoscale data storage. *Phil. Trans. Roy. Soc.* A **358**, 281.

Prinz, G. A. 1999 Magnetoelectronics. *Science* **283**, 330.

White, R. L., New, R. M. H. and Pease, R. F. W. 1997 Patterned media: a viable route to 50 Gbit/in^2 and up for magnetic recording? *IEEE Trans. Magn.* **33**, 990.

9 Future high-capacity optical communications: systems and networks

Polina Bayvel

Royal Society University Research Fellow/Reader, Department of Electronic and Electrical Engineering, University College London, Torrington Place, London WC1E 7JE, UK

9.1 Introduction

The relentless growth of telecommunications, Internet and information technologies has undoubtedly changed people's lives forever. It has allowed people from all corners of the world to share information and ideas. This revolution in technology is likely to continue to shape and alter the way we live in all facets of life such as education, health, commerce and entertainment. New network applications and new information will lead to further growth in capacity of the network. The single most important area of technology underlying the explosion in network capacity, underpinning the development of the Internet and guiding this information revolution is that of optical communications. Optical communications is a relatively recent area of research. However, it is also one that is experiencing extraordinary changes and the transfer from the research laboratory to the field has been rapid.

Since the demonstration of the first low-loss optical fibres in the 1970s, and the introduction of the first optical communication system across the Atlantic ocean, TAT-8, installed in 1988, which used electrical regenerators to amplify, re-shape and re-time the signals and operated at 140 Mbit s^{-1}, the field has witnessed a revolution, with an accelerating pace that shows no sign of saturation. In more recent years the capacity of optical networks to transmit data has increased vastly, to in excess of several terabits per fibre (tera $= 10^{12}$). Of course, this information revolution has important implications for society and the global economy in the long run. However, this issue is beyond the subject of this article. It will be enough of a challenge to describe here, in the space available, the main technical developments in the field and the unsolved research problems.

The technological advances in the area of optical communications have led to the emergence of a new area of research, one that combines many previously distinct research activities. These include fundamental research into optical devices (such as semiconductor physics and nonlinear optics), research on the transmission of electromagnetic radiation in guiding optical media and transmission systems, multiplexing technology and complex traffic and network theory.

The primary limit on the amount of data which can be carried from point to point is the available transmission bandwidth of the medium itself. Single-mode optical fibres are now the ubiquitous transmission medium for the largest part of the telecommunications network, except for the last 3 km between the subscriber and the local exchange. The realisation that the transmission bandwidth of single-mode optical fibres (as shown in Figure 9.1) can provide some 100 nm of optical bandwidth, defined by the low-loss spectral region of 1500–1600 nm, has stimulated one of the most buoyant areas of research. A wavelength range of 100 nm represents over 12 THz of potential frequency space for the transmission of data. Removal of the hydroxyl (OH^-) impurity from the silica fibre will reduce the absorption at <1500 nm, extending the available bandwidth, to over 36 THz.

What is then the best way to tap into the huge reserve of bandwidth, which exceeds all the radio, HF and satellite spectrum by several orders of magnitude?

9.2 Slicing up the optical bandwidth

Two approaches are clearly possible. Wavelength-division multiplexing or WDM, shown in Figure 9.2, allows the spectrum to be sliced up into channels with a different wavelength allocated to each data-carrying channel, similarly to the frequency-division multiplexing in radio technology. The spectral efficiency of accessing this bandwidth is clearly important since it defines how many channels can be 'slotted' within the available spectrum. Fundamentally, the spectral efficiency is determined by the availability of stable single-frequency sources and demultiplexers/routers able to select individual wavelength channels from the entire spectral comb. The closer

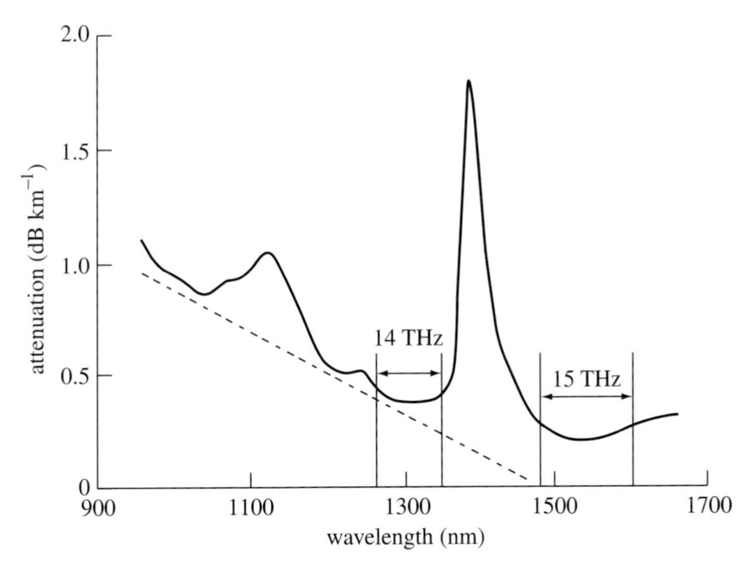

Figure 9.1. Showing the bandwidth of single-mode optical fibres defined by the regions of low attenuation; the dotted line defines the fundamental Rayleigh scattering limit arising from random fluctuations in density frozen into fused silica during fabrication of the fibre. These local fluctuations in the refractive index scatter the light in all directions; the peaks in the attenuation below 1500 nm correspond to absorption by OH^--ion impurities. Beyond 1600 nm, pure silica becomes highly absorptive (not shown).

the channels can be spaced the denser the WDM system is said to be. The more dense the system the more worthwhile the investment to develop this initially temperamental technology. The 'granularity', that is, the convenient rate of electrical modulation 2.5 Gbit s^{-1} or perhaps even as high as 40 or 80 Gbit s^{-1}, deemed high enough to be assigned an individual wavelength is the subject of some debate.

A second, competing, approach, known as optical time-division multiplexing (OTDM), shown in Figure 9.3, would involve all-optical interleaving and regeneration of many lower-bit-rate channels at 10 Gbit s^{-1} all at the same wavelength, optically multiplexed to 100 Gbit s^{-1} and over (Cotter *et al.* 1999) The interleaving would be performed by first generating different streams of very narrow pulses, say for a bit rate of 10 Gbit s^{-1} with a bit period of 100 ps, the pulse duration would be of the order of several picoseconds and each stream to be multiplexed would be separately

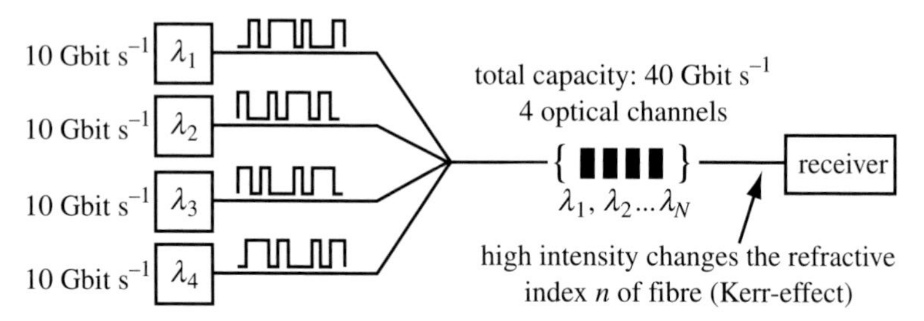

Figure 9.2. A schematic diagram of a wavelength-division multiplexing (WDM) concept, showing how, for example, use of four channels at different wavelengths in a single fibre, each individually modulated at 10 Gbit s^{-1}, can give the total capacity of 40 Gbit s^{-1}.

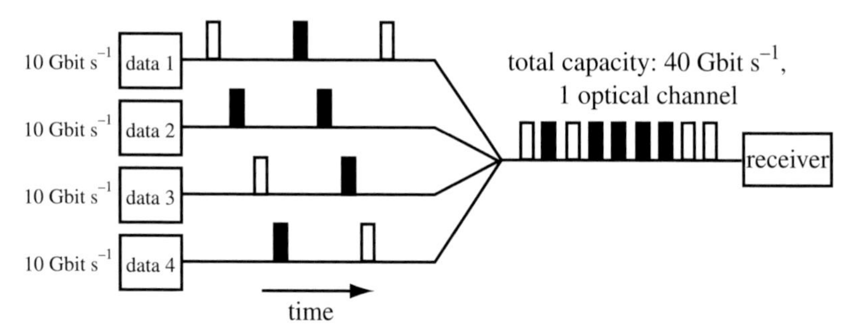

Figure 9.3. The approach used to achieve 40 Gbits s^{-1} in the fibre using a single wavelength with optical time-division multiplexing (OTDM). The total amount of data is contained in one optical channel, which is transmitted serially and interleaved in time.

modulated and delayed, using precise optical delays in free-space, optical-fibre waveguides (2 mm for a 10 ps delay in a silica fibre waveguide) with reference to the initial bit stream, generating an aggregate bit rate of $n \times 10$ Gbit s^{-1}.

Paradoxically, these transmission technologies, which are intended to increase the amount of data which can be transported from point to point, both exacerbate the 'electronic bottleneck'. This eloquently describes the limitation of the electronic switches and routers to process the very-high-bit-rate signals terminating at their ports. The telecommunication network consists of high-capacity links, connected by electronic processing nodes, switches or crossconnects. As the number of high-capacity links increases, the size of the electronic routing switch must increase too. There are fundamental limitations on this. Electronic switching or cross-connection requires the processing of signals at much lower capacity, say at 155 Mbit s^{-1}, hence any terminated signal with a higher bit rate of, for example, 10 Gbit s^{-1} requires electrical demultiplexing to sixty-four ports. The growth in electronic ports is limited by the electronic interconnect technology and by the delays in processing associated with the computational complexity of looking up the route for each 155 Mbit s^{-1} signal and repeating this for all the 10-Gbit s^{-1} channels. Much work has been done on improving the design of electronic routers by speeding up route-look-up algorithms. Yet most of these scale poorly with the size of the routing table (see, for example, Kumar *et al.*, (1998)).

So, the potential of using the wavelength domain not just for transmission but also for routing proved irresistibly appealing, spurring on the work in a large number of research laboratories. In this conceptually simple approach, termed wavelength routing, wavelengths can be used to denote routing destinations, directing high-capacity all-optical signals, transparently, between source and destination nodes without the need for electronic processing. The routing and allocation of wavelength would then be pre-determined. Several questions immediately arise.

- How many wavelengths would be required to provide the required connectivity without intermediate opto-electronic conversion or switching? The answer to this question defines the density of the channel spacing, the required stability of the laser sources and the accuracy of the wavelength-selective components (lasers, demultiplexers, filters and crossconnects).
- Would some form of all-optical wavelength conversion be required in order to avoid contention for wavelengths?
- How many wavelengths could successfully be transmitted (and over which distances) without incurring errors through interactions among channels caused by a combination of nonlinearities of fibres at high optical intensities (since even the modest optical powers of several milliwatts per channel in single-mode optical fibres with core radii of 8–10 μm give rise to huge optical intensities of the order of 100 MW m^{-2}!), chromatic dispersion of the wavelength comb and the accumulated crosstalk from imperfect selectivity of filters?
- How could an all-optical network be managed? After all, only four parameters would be available for monitoring, namely power, signal-to-noise ratio, spectrum and bit-error-rate instead of a panoply of electronic monitoring and error correction.

Most of the potential limitations are associated with the fundamentally analogue nature of WDM and the proponents of OTDM were quick to point out that this approach naturally lends itself to digital processing techniques, including digital signal regeneration and bit-serial processing. Both short-pulse generation (using for example an external cavity mode-locked laser) and regeneration using nonlinear effects in fibres or semiconductors are the subjects of much intensive research, which is at a rather early stage, and this approach has very recently seen something of a surge (Cotter *et al.* 1999). It offers the potential of a greatly simplified design of optical amplifier and management of dispersion due to single-wavelength transmission. The extension of this transmission technique to network-wide applications has also a number of associated research issues:

- Devices for generation of stable, wavelength-tunable ultra-narrow optical pulses are needed.
- Transmission of short pulses even over medium-haul distances (100–1000 km) requires careful optimisation of fibre dispersion, source wavelength, polarisation and channel delays for correct sub-rate interleaving.
- In the context of extending these techniques to optical packet networks, how can an essentially non-optical functionality of buffering or memory be implemented optically, given the absence of optical random-access memory?
- Synchronisation of packets at nodes is needed.

9.3 The evolution of WDM ideas: from concept to reality

As mentioned before, the concepts of WDM were first proposed by analogy with an 'optical ether' or frequency-division multiplexing over fifteen years ago, but the ideas languished for many years until the early 1990s. The enormous growth in demand for high-capacity links over substantial distances of several thousand kilometres resulted in a rapid exhaustion of the optical-fibre networks that had been installed. Since the installation of optical-fibre cable is extremely time-consuming and expensive, alternative approaches to expanding the installed capacity became attractive. The early ideas were 'rediscovered', the impetus being provided by the development of more precise optical components; namely more stable lasers and all-optical erbium-doped fibre amplifiers (which could now replace electri-

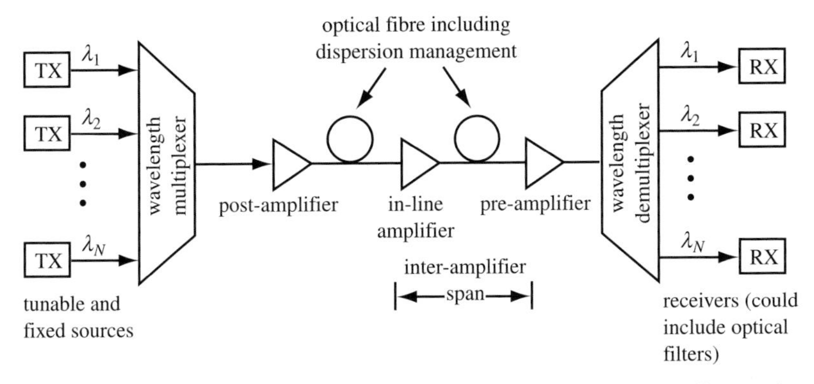

Figure 9.4. Key building blocks of a WDM transmission system using all-optical amplifiers and N channels of different wavelengths (TX denotes a transmitter and RX denotes a receiver).

cal regenerators to amplify the entire spectral comb with a single optically pumped amplifier) and the emergence from research laboratories of the first practical, compact and stable optical filters that could operate as demultiplexers (see Figure 9.4 illustrating a typical WDM system). Although initially the research work separated into two distinct and mutually ambivalent camps of proponents of 'linear' WDM networks and 'nonlinear' soliton systems, these have now begun to converge.

In a linear or non-return-to-zero (NRZ) modulation format (shown in Figure 9.5), the optical intensity is turned on and off with the bit period equal to the inverse of the rate of electrical modulation or bit rate. Any appreciable dispersion in the fibre (due to the wavelength-dependent mode shape and variation of the refractive index) leads to pulse spreading and the spilling-over of energy into the following bit period, resulting in transmission penalties. Dispersion-shifted fibres with near zero dispersion are optimal for single-channel transmission because they allow one to minimise the linear dispersion-induced pulse broadening. A second distorting effect is the nonlinear phase shift that builds up on transmission of a high-intensity channel as a result of the change in refractive index with optical intensity due to the Kerr nonlinearity in the fibre, which is known as self-phase modulation and can be minimised in low-dispersion fibres. However, the use of low-dispersion fibres is highly detrimental in multi-wavelength transmission, because the combination of high optical intensities and low dispersion leads to phase matching between the different

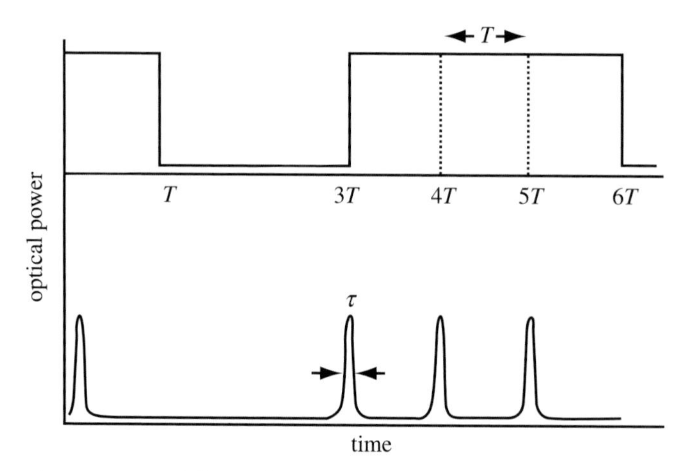

Figure 9.5. Two different formats for optical modulation, non-return-to-zero (NRZ) and return-to-zero (RZ). T is the bit perod, equal to the reciprocal of the bit rate, and the ratio τ/T could be 0.2–0.5, depending on the system. The RZ format requires optical sources capable of generating very short pulses.

wavelength channels, in turn leading to generation of new 'unwanted' frequency components (an effect known as 'four-wave mixing' (FWM)). Even more detrimentally, the phase modulation of one channel by its neighbours, which can be converted into intensity modulation in the course of transmission, is a result of another unwanted nonlinearity, namely cross-phase modulation (XPM). The investigation of the relative magnitudes and importance of these in various system configurations with various numbers of channels and channel spacings is the subject of much buoyant research in many laboratories around the world. Economics points to the necessary increase in the distance between amplifiers. This in turn leads to a higher power per channel being required to maintain the signal-to-noise ratio for each channel in order to ensure that the bit-error rate is low. High channel powers inevitably lead to impairments of transmission due to nonlinearities of fibres, the interplay between which must first be understood and then minimised by appropriate tailoring of the dispersion in the transmission fibre, amplifier design and choice of channel spacing.

The competing technology, the transmission of return-to-zero (RZ) pulses, also shown in Figure 9.5, whose duration is much less than the bit period, say by a factor of 3–5, was initially considered a solution to the problem of nonlinear interactions and specifically self-phase-modulation-induced distortion. These pulses, termed 'solitons', balance the effects of self-phase modulation and dispersion in the fibre to allow long-distance transmission without pulse broadening. RZ-pulse generation itself is the subject of a considerable amount of research activity and RZ pulses are a pre-requisite for OTDM transmission since several pulses must be temporally interleaved within the bit period to generate a much higher aggregate signal, as has already been mentioned. In actual fact, RZ pulses may, but need not evolve into soliton pulses, depending on their energies and the total distance of transmission, which must be sufficiently long to prove that the pulse is indeed a soliton, i.e. one that does not spread in time or frequency on transmission, independently of the distance travelled.

Both NRZ and RZ pulses are amenable to wavelength multiplexing, but the RZ format is obviously more spectrally demanding, given that it employs much shorter initial pulses that require greater channel spacing.

9.4 Competition for the highest bit-rate–distance product

Leading research laboratories around the world have competed to report the longest, and ever increasing, transmission distances with the highest bit rate × distance product through cleverly optimised fibre dispersion, channel spacing and inter-amplifier distance. Much of the competition has been between North American and Japanese research laboratories, and some of the key ground-breaking experimental results are listed in Table 9.1, although many records are likely to be exceeded by the time this article is published!

The first really multi-wavelength experiments were reported in 1995 of 20 × 5 Gbit s^{-1} over 6000 km by Bell Laboratories (Bergano & Davidson 1995) to evolve in 1999 to 64 channels × 10 Gbit s^{-1} over 7200 km spaced by (0.24 nm) 30 GHz (Bergano *et al.* 1999). The very latest results from this group have demonstrated over 12 petabits s^{-1} per km (peta = 10^{15}) with 180 × 10 Gbit s^{-1} channels over 7000 km with 0.21 nm channel spacing and ultrawide (43 nm) optical amplifiers. Longer-distance transmission has also been reported in ground-breaking research experiments by groups in Japan at KDD for transoceanic distances and NTT Research Laboratories for shorter distances in experiments. Longer-distance transmission has also been reported through 100 × 10.7 Gbit s^{-1} transmission over 6200–10 000 km with 0.3–0.4 nm channel spacing (Tsuritani 1999, Naito 1999). At the high bit rate end 'hero' experiments have exceeded 3 Tbit s^{-1}

WDM with the demonstration of 19×160 Gbit s^{-1} OTDM channels over 40 km fibre by Kawanishi (1999) just beating the previous record of 7×200 Gbit s^{-1} channels over 50 km (Kawanishi 1997) and 13×80 Gbit s^{-1} OTDM channels spaced by 200 GHz over 89 km (Miyamoto 1998). Highest single channel (by optical time interleaving of 64×10 Gbit s^{-1} modulated pulses and polarisation multiplexing) transmission of 1.28 Tbit s^{-1} over 70 km was shown by Nakazawa (2000).

Table 9.1 also highlights the conflict between high bit rate per channel and achievable transmission distances. It is clear that the fight for the longest bit rate $\times$ distance product has been won in long-distance transmission experiments exceeding 15 petabit s^{-1} per km in 2000. The largest bandwidth (that is number of channel $\times$ bit rate per channel) is approximately 7 Tbit s^{-1} with a record spectral efficiency of 0.8 bit s^{-1} per Hz with the competing research labs squeezing out the last wavelength channel to achieve the results. Achievable distances are reduced with higher bit rate per channel even though these experiments use some form of nonlinear supported transmission of RZ pulses to offset the effects of nonlinearities and dispersion as well as complex dispersion compensation, including dispersion slope compensation requiring special fibre design. This is also the subject of much intensive research all over the world.

9.5 Wavelength-routed optical networks

As has already been mentioned, the use of wavelength for network-wide routing is an appealing approach. It is an extension of point-to-point WDM, since the number of wavelengths required is determined not only by impairments of transmission but also by the demand for traffic and the physical connectivity of the network. A wavelength-routed optical network (WRON) will consist of the interconnection of a large number of point-to-point WDM systems by optical routers or crossconnects able to route channels according to their wavelength, 'transparently', so to speak. This has the potential of eliminating the electronic processing at network nodes which is one of the key limitations on the growth of bandwidth. A key question, therefore, concerns exactly how many wavelengths would be required to interconnect network nodes on a national or world-wide scale to satisfy a given demand for traffic. Research aimed at finding the answer to this seemingly straightforward question has generated a huge number of research publications all around the world.

Table 9.1. *Bit-rate × distance products for some recent record single- and multi-channel experiments: the experiments are arbitrarily listed in order of bit rate per channel. However, researchers claim different record values – highest channel or aggregate bit rate per fibre (for the former (Nakazawa* et al.*) with 640 Gbit* s^{-1} *for the latter (Faerbert* et al.*) 7 Tbit* s^{-1}*) bit rate distance products, total number of channels (both belong to Tanaka* et al. *– 15 petabit* s^{-1} *per km, longest total transmission distance . . . Most of the dramatic experiments over the years have been reported at the two key conferences in the field: ECOC – European Conference on Optical Communications (typically held in September, since 1974) and OFC – Optical Fibre Communications Conference, typically held in February/March, since 1975)*

Author	Number of channels	Bit rate (Gbits s^{-1})	Distance (km)	Bit rate × distance (Tbits s^{-1} per km)
Bergano (ECOC 95)	20	5	6000	600
Taga (OFC 98)	20	10.7	9000	1990
Murakami (ECOC 98)	25	10	9300	2325
Bergano (OFC 99)	64	10.7	7200	4931
Tsuritani (ECOC 99)	100	10.7	6200	7270
Naito (ECOC 99)	104	10	10127	10532
Davidson (OFC 2000)	180	10	7000	12600
Tanaka (ECOC 2000)	211	10	7221	15236
Yano (ECOC 96)	132	20	120	317
Ito (OFC 2000)	160	20	1500	4800
Nielsen (OFC 2000)	82	40	300	984
Bigo (ECOC 2000)	128	40	300	1536
Ito (ECOC 2000)	160	40	186	1190
Faerbert (ECOC 2000)	176	40	50	352
Miyamoto (ECOC 98)	13	80	89	93
Kawanishi (OFC 99)	19	160	40	121
Raybon (OFC 2000)	1	320	200	64
Nakazawa (ECOC 2000)	1	1280	70	90

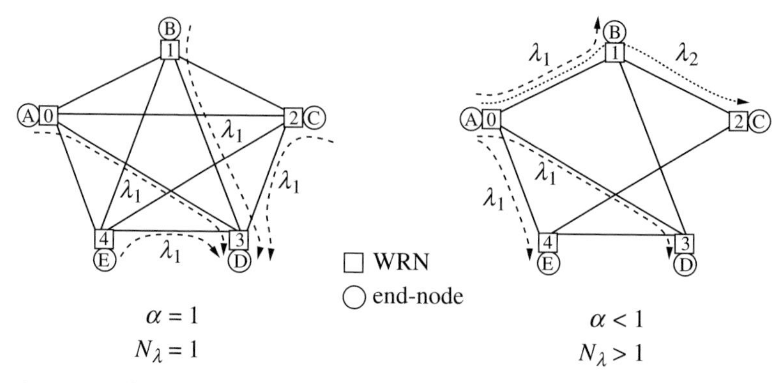

Figure 9.6. The architecture for wavelength-routed optical networks (WRONs). In this five-node example, ten logical connections must be mapped onto a fully connected network. The connectivity, α, is defined as $2L/[N(N-1)]$, where L is the number of links in the network and N is the number of nodes. In the case of a fully connected network, only one wavelength is necessary ($N_\lambda = 1$), whilst a sparsely connected network needs more wavelength to avoid a clash of colours. How many wavelengths are necessary to connect a network with an arbitrary physical connectivity?

The question is trickier than it at first appears and, yet, the correct answer determines both the network architecture and the requirements on devices – such as the channel spacing, designs of optical amplifiers and routers and many details of the system and of the network. The WRON approach requires a conceptually different approach to the design and operation of telecommunications networks.

To illustrate the difficulty of the task, consider the problem in Figure 9.6, which shows a simple five-node network fully interconnected by bi-directional optical-fibre links with each node connected to four others, so that each node has a unique physical path, disjoint from those for other pairs of nodes. Each node consists of a transmitter-and-receiver array and a wavelength-routing node. The transmitters/receivers emit and receive wavelengths corresponding to given lightpaths. Assuming that each node needs to communicate to all the others, in a five-node network ten connections need to be made. The same wavelength could then be used for all the connections. However, in a real physical network the connectivity is far from full and the required logical connections must be mapped onto the available physical connectivity, without different channels with the same wavelengths being transmitted simultaneously in the same physical link,

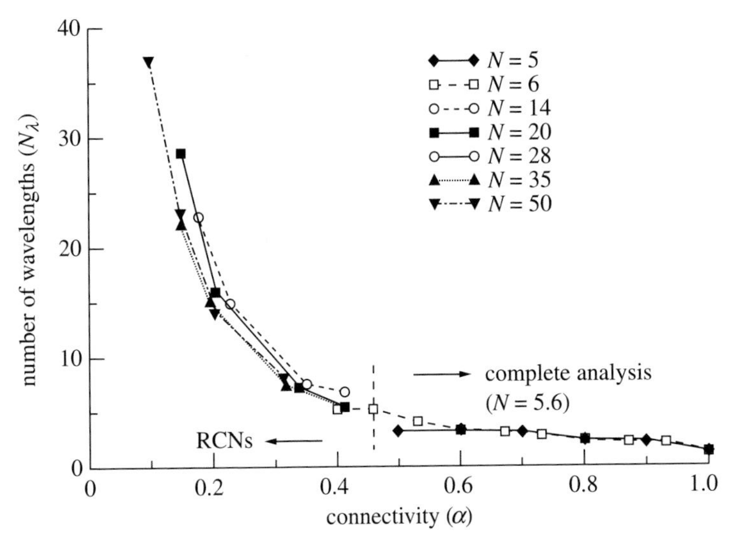

Figure 9.7. The average number of wavelengths against the physical connectivity for networks with various numbers of nodes (from Baroni and Bayvel (1997)). It is interesting to note that the number of wavelengths required varies with the connectivity and is almost independent of the number of nodes. 'RCN' denotes randomly connected networks; namely networks with arbitrary connectivity, for which only a small number of all possible topologies can be analysed, whereas the complete-analysis part of the graph refers to small networks with five and six nodes, whose topologies can be analysed completely. Note that, for a connectivity of 1, that is a fully connected network, only one wavelength is required (as in Figure 9.6).

which is referred to as contention for wavelengths. Moreover, the actual physical connectivity is far from regular and the calculation of the number of wavelengths required for a connectivity of less than 100 per cent is the key issue (see Figure 9.7 showing results for arbitrarily connected networks). The difficulty arises since, given a certain number of links, lots of different physical topologies can be constructed, each requiring a different number of wavelengths. Ideally one would aim to design a network needing the fewest wavelengths. The problem of analysis is converted into that of synthesis of good topologies.

An even more difficult question is that of what happens when links fail – each link would now possibly be carrying a large number of wavelength channels or lightpaths that would have to be restored by re-routing them onto different, surviving, links. How should these lightpaths best be re-routed to require the least number of extra wavelengths without contention? Should the restoration technique simply by-pass the failed link or should a signal be sent to the originating node transmitters to re-route all the lightpaths along pre-calculated alternative paths? Another group of sub-problems is related to the case of the number of wavelengths per fibre being fixed and the problem was related to minimising the total number of fibres for a given topology. In both cases, the question of adding growth of traffic to existing lightpaths complicates the analysis.

The problem of allocation of wavelengths and/or routing pre-occupied many groups of researchers in the new field of optical network design which combined complex techniques borrowed from discrete mathematics, graph theory and combinatorics with the understanding of the principles of optical communication (see for example Baroni and Bayvel (1997) and Stern and Bala (1999)). In many cases the fomulations of the problem were too complex to be solved analytically or by using integer programming methods and, hence, formulation of heuristic (or automated rule-of-thumb) algorithms was required.

Associated with the questions concerning the design of networks were questions about how best to design and control an optical crossconnect or router. For example, should they be fixed or reconfigurable (allowing a wavelength to be changed along its route)? A battle in the literature and at conferences concerned the necessity of wavelength conversion using devices that would change the wavelength of a given lightpath along the path if contention for wavelength occurred. Proponents argued that the use of wavelength converters would make the allocation of wavelengths more flexible, ignoring the fact that one per wavelength per fibre would be required, resulting in a large number of wavelength-converter devices needing monitoring and control. Should these devices be all-optical, based on using nonlinear effects in semiconductors to generate new frequency, or opto-electronic, working simply by detecting the signal electrically and then re-transmitting it at a different wavelength (somewhat defeating approach of the all-optical routing!)?

Despite the buzz of activity in this area, the difficulty of achieving consensus has slowed the implementation of these networks. The real difficulty can be traced to the absence of a good reliable optical switch that can be scaled to a large number of ports.

Several techniques are nominally possible, yet none has been shown to give reliable operation for thousands of ports. Research papers have reported 32×32 port routers, far short of what is required. So why the difficulty? The story of optical switching has long suffered from an undermining of credibility, dating to the days of the hyped-up all-optical computing. Despite the recent dramatic advances in optical technology, optical logic devices are still relatively primitive compared with electronics: they are relatively bulky and poorly integrated, have limited cascadability and high losses and are power-hungry and, surprisingly, relatively slow (slower than the achievable line rates). In general, an optical wavelength router consists of two functions – a diffractive element (a demultiplexer) that separates the wavelength channels and an array of logical on–off switches, with a high contrast ratio to process these. A high constrast (on–off) ratio is vital in order to minimise the crosstalk or leaking of light at unwanted wavelengths when these devices are cascades. Whilst the problem of the diffractive function has effectively been solved using either low-crosstalk and high-resolution free-space grating devices or planar wavelength-selective waveguide-grating router arrays, an efficient switching functionality is still some way off.

One possible approach that has shown some promise is a switch consisting of an array of tiny, perhaps 20 μm × 10 μm, micro-machined devices, collectively termed MEMS (Micro-electro-mechanical system), made of silica, acting as a two-dimensional array of paddles either transmitting or blocking the beam (as shown in Figure 9.8). This technology is potentially simple and wavelength-independent with a high contrast ratio, but it appears fundamentally unscalable beyond tens of ports, since light beams propagating in free space through the array of paddles must be re-collimated periodically and, because the number of paddles passed varies depending on the route, the problem of collimation is a challenging one. However, all MEMS devices are relatively slow (compared with line bit rates of tens of Gbit s^{-1}) with maximum speeds probably of fractions of milliseconds.

However, recently much interest has been shown in the reflective MEMS arrays which do not require complex collimation and consist of two-axis beam-steering micromechanical mirrors used to reflect light beams for routing between fibres (Neilson *et al.* 2000).

An alternative approach would be to use semiconductor optical amplifiers in a nonlinear interferometric configuration capable of operation at speeds of 100 Gbit s^{-1} as described by Cotter *et al.* (1999). In this device, the input signal is split between two arms of an interferometer. The

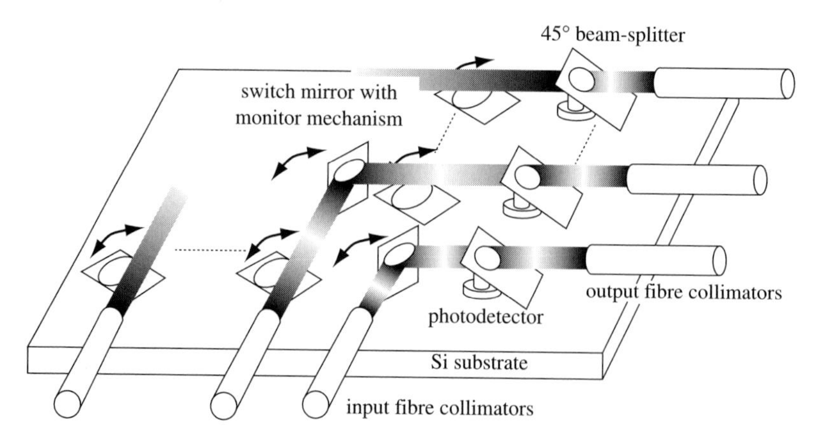

Figure 9.8. An example of a free-space micro-machined optical switching matrix made from silicon (from Lin *et al.* (1998)). The switching mirror includes an operation-monitoring mechanism that monitors the configuration of the switching fabric. The photodetector detects the states and quality of signals and verifies the switching path.

interometer is balanced so that, in the absence of a control signal, the input signal emerges from one output port (Figure 9.9). The effect of applyng the control signal is to induce a differential phase shift between the two arms so that the input signal is switched over to a second output port. It is not known whether these devices can be integrated, but they offer promise for the future.

Ultimately the main advantage of WRONs, namely their ability to interconnect high-capacity pipes in the top layer of the network, has been questioned by the re-emergence of the packet networks. The new challenge is to understand how packets, rather than continuous high-capacity signals, can be routed across the telecommunications network.

9.6 Packet networks: the ultimate application of optical networking?

WRONs and high-capacity WDM transmission are the solution to the electronic bottleneck in the transport network. Electrically multiplexed signals are composed of millions of lower-speed voice and data signals

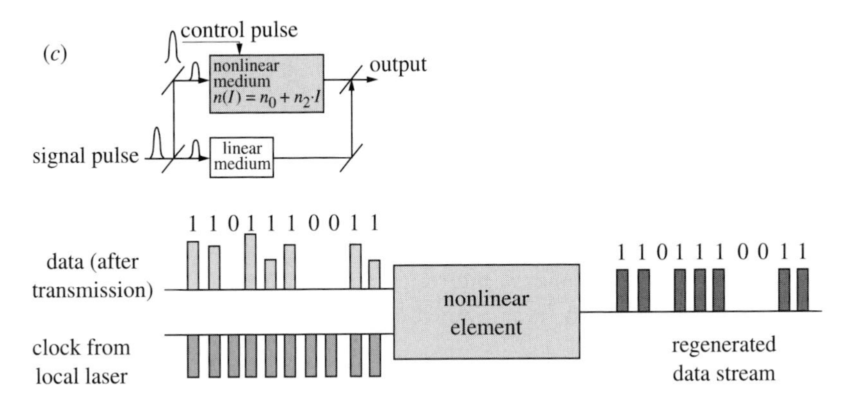

Figure 9.9. An interferometric nonlinear switching element used as a gate for switching and regeneration of packets.

before being transported in high-capacity pipes. There are a lot of 'layers' between the user's layer and the capacity-rich optical transmission layer. The reason for this is that the high-capacity networks serve millions of users and the cost per user is relatively modest due to this sharing. However, the cost of providing a high bandwidth to the individual user would be great and how could the provider ever see a return on the investment? The resources at the user's end are likely to be limited for the foreseeable future. One way to help solve the mismatch between access and transport is by achieving a better utilisation of available resources through use of packet networks. In this way, any transmission is not continuous, thus tying up the entire channel, but rather is split up into smaller units or packets, of either variable or fixed length. Each packet has an address of the destination node and a sequence number and is sent into the network to be routed by a number of intermediate packet routers using the addressing information in each packet. At the intermediate or transit routers, there may be contention among packets and hence there is a requirement for delays or buffering. At the destination node the original data are re-assembled from sequenced packets. If there are too many packets for the routers, buffers or packets become damaged on transmission through errors or noise; they are said to be lost and may have to be re-sent. The Internet is an example of a packet-switching network.

What are packets? They are, in fact, simply bursts of 1s and 0s of a predetermined length, say 3000–13000 bits with a header containing the packet's destination address of, say, up to 10 per cent of the payload (for example 300 bits). The packet rate, i.e. the number of packets transmitted per second, is the transmission channel rate divided by the number of bits per packet. Each packet could be carrying a different number of bits at a different internal (i.e. specific to the packet) bit rate. Small packets can be combined into larger packets and routed between users first in the electronic domain and then in an optical domain. Each channel can then be used more efficiently through better multiplexing or fitting of packets into the available capacity. It is proposed that, just like for the electronic packet switching, at the user's end of the networks, optical packet networks can help in the ultimate quest for transparency by providing optical equivalents of electronic packet networks, namely buffering, synchronisation and routing functions including recognition and processing of the header, that could be shown to be real and to offer advantages.

In terms of the earlier discussion on WDM versus OTDM, packets could be transmitted either at high single-wavelength or at multi-wavelength OTDM rates, sequentially or in parallel using different wavelengths. In fact, bit-interleaved TDM systems are isomorphic to WDM. The relative merits of the two approaches have yet to be determined and the means of generation of optical packets from electrical tributaries representing different services and requiring different qualities of service is far from defined. Yet the general area is being seen as an exciting one filled with demanding problems requiring research.

The functionality of an optical packet network, independently of the exact implementation technology, must nonetheless include several key components. The most important is buffering – the ability to delay packets during recognition and processing of the header and in the process of switching to avoid contention among packets. Since truly random-access memory does not exist in optics, the only way to introduce buffering is through the use of precisely set optical delays – typically involving the use of a recirculating fibre delay line on which a packet can be delayed by a multiple of the round-trip time or – which is functionally the same – an array of waveguides of different lengths. To date no one has yet proposed a solution to the problem of the finite memory-clearance time – since this will be at least as long as a delay itself! However, the higher the channel

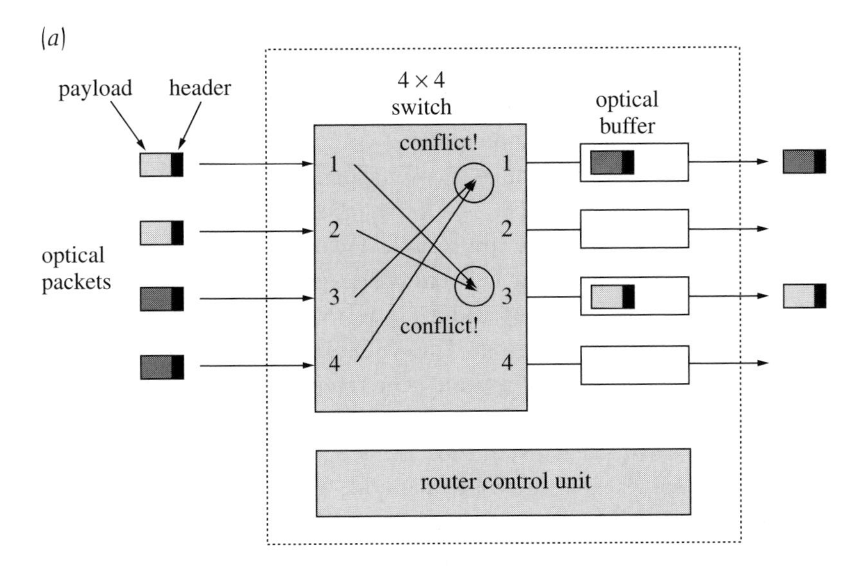

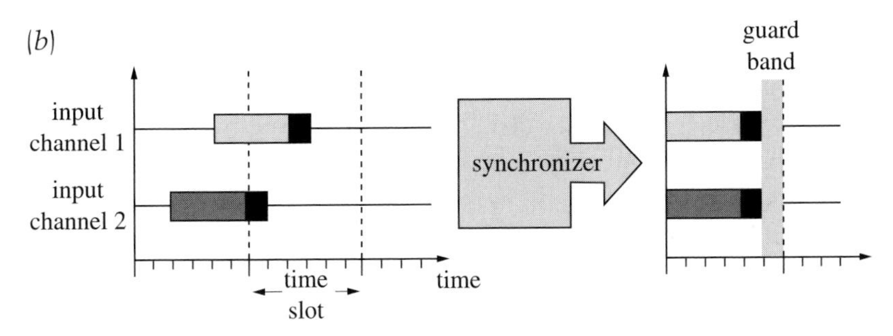

Figure 9.10. (a) Optical delay and buffering functions required to avoid contention among packets. Conflicts arise when packets seek access to the same output in the same time slot and can be solved by buffering and delaying packets. (b) Synchronisation of packets is needed to prevent packets being 'chopped up' at nodes.

rate the shorter the buffers need to be and the more practical the network becomes. It is clear that, despite much promise and the awsome advantages optical networks would bring to satisfying growing demands on networks, the photonic-network designer will be forced to invent new network architectures and protocols, rather than simply imitating and scaling up designs developed by electronic engineers. For example, new strategies will be needed to resolve contention (which arises when two packets of data are routed towards the same pipe simultaneously, as shown in Figure 9.10, and can be solved in electronics easily by using electronic buffers) in a way that overcomes the lack of optical memory.

Another key function, which is again applicable both to WDM- and to OTDM-based packet networks, is address recognition. In electronic networks, addresses are loaded into a register and compared with a stored bit pattern. For very fast photonic networks this is problematic. One proposed solution is to implement an optical AND gate where a packet is separated from its an optical-routing header. The synchronisation pulse, followed by the address of each packet, separated from the payload, is compared with the keyword set for each node by a series of pre-fixed delays. The synchronisation is crucial since each node must be able to recognise the start of the packet and, assuming that there is no network-wide clock (which is almost impossible to implement), so is some sort of asynchronous operation whereby timing (and thus alignment of packets and all processing) must be carried out at the packet rate rather than at the bit level. If there is a match, the output of the AND gate triggers the switching action. Further delays/buffering must be introduced in order to avoid any contention. One rather ambitious proposal to avoid the need for memory or buffering has been that of deflection or 'hot-potato' routing whereby, in the case of contention among packets, the packet will be deflected and sent from node to node in a nomadic fashion until it can reach its destination. It is far from clear that this is realistically implementable without incurring the penalty of prohibitive delays.

The clear perceived advantage with OTDM is the possiblity of regeneration, demultiplexing and switching. All these functions can be realised using nonlinearities of the material, either in semiconductors or in a fibre, by implementing interferometric switches whose refractive indices are dependent on intensity. These can be made to operate as simple logic gates. Whether these will be implemented in semiconductors or as optical fibre-based devices is not yet known. Whilst they are functionally the same, their time responses are different – semiconductors are compact but limited in speed to picosecond operation whereas nonlinear fibre devices are relatively long (tens of metres) but have significantly faster switching speeds (in the sub-picosecond range). To date all-optical regeneration at 40 Gbits s^{-1} and higher, using an ultra-fast nonlinear interferometer, demultiplexing to 10 Gbits s^{-1} in a nonlinear optical loop mirror from a myriad of rates up to 1.28 Tbit s^{-1} and packet recognition at rates of up to 100 Gbits s^{-1} have been achieved. The implementation of a network consisting of these building blocks is one of the challenges for the future.

9.7 Future battles for bandwidth

The bandwidth in the optical domain, unlike in almost any other part of the radio and microwave spectrum, is 'up for grabs'. There are numerous new optical devices and almost infinite transmission capacity, yet the best way of using it, whether through time-division or wavelength-division multiplexing, processing it as packets or in transparent high-capacity pipes, is far from clear. Curiously, even researchers in the field could not imagine the rate at which the bandwidth is being consumed and how bandwidth-hungry the global population has become and how capable of generating unimaginable quantities of data. Although we must always question the implication for society of this expansion in quantity of information, the challenge for the next millennium is to design conceptually new networks that can carry it as far, and as fast, as possible, and process it flexibly, quickly and reliably.

9.8 Further reading

Some of the following may give further insight and the list of references for Table 9.1 is partially contained in Bayvel (2000) and latest references for 2000 are from the *Technical Digest of the 25th Optical Fiber Communications Conference*, held during March 2000, in Baltimore, Maryland, USA and the *Proceedings of the 26th European Conference on Optical Communications*, held in September 2000 in Munich, Germany.

Baroni, S. and Bayvel, P. 1997 Wavelength requirements in arbitrarily connected wavelength routed optical networks. *IEEE J. Lightwave Technol.* **15**, 242.

Baroni, S., Bayvel, P., Gibbens, R. J. and Korotky, S. K. 1999 Design of resilient multifibre wavelength-routed optical transport networks. *IEEE J. Lightwave Technol.* **17**, 743.

Bayvel, P., 2000 Future high capacity optical telecommunication networks. *Phil Trans. Roy. Soc.* A **358**, 303–29.

Bayvel, P., Gibbens, R. J. and Midwinter, J. E. (editors) 2000 Network modelling in the 21st century. *Phil Trans Roy Soc* A. Discussion meeting issue, **352**, August 2000.

Cotter, D., Manning, R.J., Blow, K. J., Ellis, A. D., Kelly, A. E., Nesset, D., Phillips, I. D., Poustie, A. J. and Rodgers, D. C. 1999 Nonlinear optics for high-speed digital information processing. *Science* **286**, 1523–8.

Kaminow, I. P. and Koch, T. L. (editors) 1997 *Optical Fiber Telecommunications* volumes IIIA and IIIB. New York: Academic Press.

Kumar V. P., Lakshman, T. V. and Stiliadis, D. 1998 Beyond best effort: router architectures for the differentiated services of tomorrow's internet. *IEEE Communications Magazine* May, 152–64.

Lagasse, P. *et al.* 1998 *Photonic Technologies in Europe* edited by Descamps, C. Norway: Telenor AS R&D. Thematic issues from ACTS, http://www.inforwin.org.

Lin, L. Y., Lunardi, L. M., and Goldstein, E. L. 1998 Optical cross-connect integrated systems: a free-space micromachined module for signal and switching configuration monitoring. In *Proc. IEEE/LEOS Summer Topical Meeting on 'Optical MEMs'*, paper PD004.

Stern, T. E. and Bala, K., 1999 *Multiwavelength Optical Networks*. Reading, MA: Addison Wesley.

10
Control and pricing of the Internet

Richard Gibbens

Statistical Laboratory, Centre for Mathematical Sciences, University of Cambridge, Wilberforce Road, Cambridge CB3 0BW, UK

10.1 Introduction

The rapid pace of developments in communication networks has produced outstanding advances. Two parallel developments have been taking place. First, significant advances in the engineering of communication equipment, that is the switches, routers, optical fibres and all the software that coordinates and controls them, have led to a huge expansion in the bandwidths available in transmission networks. Secondly, developments in the computing technology used by the end systems have brought about a keen interest in multi-media applications as well as distributed computing; the growth of the World Wide Web and the Internet being amongst the most familiar examples of these developments.

Furthermore, such developments have started to produce major changes in the structure of many industries and in the way in which commerce is taking place. These changes will continue as the cost of access to all kinds of information continues to tumble.

All these exciting developments result from research and developments across a diverse range of fields, including mathematics, engineering, computer science and economics. This survey article, written for a general scientific audience, presents some examples of these recent advances and then develops a theme that has recently started to emerge (Gibbens and Kelly 1999a, b), which has the potential to transform the way in which we think about building communication networks in the future.

This approach is based on the view that the network need only convey feedback signals to users, perhaps in the form of marks attached to users' packets, so as to indicate the cost, measured in terms of resources consumed within the network. If a small charge is then incurred for each mark received, users would have the incentive and the necessary information to adapt their demands on traffic in such a way as to share the available resources efficiently and fairly. By speculating about the future, the article describes ways in which this approach, if it were adopted, would lead to simple and robust mechanisms for adding more elaborate notions of quality of service. Examples in which it is possible to synthesise higher-level service models associated with real-time services, such as telephony and video, from simple underlying packet networks will be given.

10.2. The current Internet

We begin our account of the current Internet by summarising its basic components and properties. We shall then look at how it has been used and what mechanisms have been developed to support these uses. Any discussion of the development of the Internet will also need to consider its funding and the means by which revenue is extracted from the users. This section will then conclude with some of the fundamental difficulties that need to be overcome before further growth in the use of the Internet can succeed.

10.2.1 The basics of packet-switched networks

The Internet is an example of a *packet-switched* network. The data to be transferred between two parties connected to such a network are not sent in one whole unit but are instead chopped up into smaller individual units called *packets*. Each separate packet is labelled with the source and destination addresses and is then dispatched between the end parties. Packets are transferred between the end parties along communication channels between intermediate network components (called *routers*) that perform routing (or switching) functions and multiplex together slower individual communication channels onto faster channels. In this way, packets traverse a network between any two end parties. All the routing functionality is performed packet by packet using the addressing information that has been attached to the data within each packet. At the routers, packets may be delayed in buffers before they can be forwarded along the next stage. When there are too many packets for the routers to buffer, packets are dropped and do not succeed in reaching their destination.

This packet-switching approach adds several types of overhead. The first is the addressing information that is added to each of the packets and the second is the additional processing load that is required at the components of the routing network in order to forward each packet. Nevertheless, this packet-switching design emerged for the Internet since it is very simple and robust. The primary aims of the design were to provide *connectivity* and *resilience* against channel or router failures. Issues of *efficiency* and *quality of service* (measured in terms of delays or dropping of packets) were of secondary importance to the design.

This idea of sending packets into the network at one point and receiving at another point those that had not been lost at intermediate congested components along the way is the lowest-level functionality of the Internet. This *best-effort* approach had the simplicity to allow many forms of communications to be built upon it. This basic functionality is provided by the protocol known as the *universal datagram protocol.*

One development was to provide a means for the end parties to monitor whether any packets had been lost and re-send them accordingly. This detection of packet loss was implemented by the sending party attaching sequence numbers together with the addressing information to the packets within a given connection. The receiving party was then required to send back packets of its own acknowledging the receipt of packets, whereby each packet could be identified by means of its sequence number. The lack of an acknowledgement by the sender within a certain time-out period was then deemed to signify that a particular packet had been lost and should therefore be re-sent. The net effect was to provide an error-free end-to-end communication channel. This is the essence of how the *transmission-control protocol* operates.

Several points are worth noting. First, the *transmission control protocol* has to establish a connection between the two end parties *before* any actual data packets are transferred (a signalling protocol exchanging information is used to *open* a connection and, later, when the connection is no longer required, to *close* a connection). Secondly, the *transmission-control protocol* is entirely implemented within the end parties' networking software. In particular, it requires no support from the network routers either in the form of additional processing of packets or in terms of storing state information about the fate of packets according to the many connections that may pass through any given router.

10.2.2 Controlling flow

Early applications of the Internet protocols included the exchange of e-mails and files using protocols, known as the *simple mail transfer protocol* and *file-transfer protocol*, respectively, built on top of the *transmission-control-protocol* connections. Congestion was soon a noticeable problem and at the heart of this issue lay the problem of the sending parties needing to be informed about the rate at which they should be sending in order to share the resources without needing to re-send too many packets. A flow-control strategy for the *transmission-control protocol* (Jacobson 1998) that used the discovery of dropping of packets to trigger the sender to reduce its sending rate was devised. The sending party then slowly increased its sending rate again until further losses of packets caused it to reduce the rate and the cycle of a slow increase and a rapid decrease was repeated. In this way, the sending parties, which share common congested network resources, had a mechanism by means of which to adapt their sending rates to balance the conflict between wishing to achieve a high throughput of packets and minimising the rate of loss of packets.

Additionally, a mechanism is provided at the start of a connection to allow the sender to increase its sending rate *rapidly* prior to beginning the cyclical phase of slowly increasing and rapidly decreasing. Strangely, this became known as the *slow-start* mechanism. Again, it is worth noting that this flow-control mechanism could be implemented entirely by the end parties and did not require the addition of any extra functionality to the network routers.

10.2.3 Applications

Many other applications that operate using the Internet to exchange information have since been developed. The most familiar application today is the World Wide Web, which provides distributed access to information that can be displayed by means of browser software. The browser software running on a client end system makes a *transmission-control-protocol* connection to a server process running on another end system and thereby exchanges the information to render a document on the browser's display. Often, a document has embedded references to other documents that are retrieved by the client in the same fashion. Thus, a single request to display a page might produce numerous *transmission-control-protocol* connections to be opened with an arbitrary collection of servers. This simple architecture has proved highly effective at allowing producers and consumers of rich multi-media information to flourish.

Another application that has received much attention recently is that of Internet telephony. This application sends and receives packets between two end parties that encode speech information. For this to provide real-time speech between two humans, the packet-sending rates should not be too low otherwise the conversation breaks up. Thus, this application is usually implemented not by using *transmission-control-protocol* connections with their associated flow-control mechanisms, but simply by means of a *Universal datagram protocol* connection. The senders do *not* reduce their rates to share the other users' demands on the common, scarce, network resources.

Much of the attention on Internet telephony has focused on issues concerning cost and quality. Since Internet-telephony calls avoid parts of the public switched-telephony network, they also avoid their charges. However, the quality of the connections rarely matches that of the public switched-telephony networks due to the impact of the packet-loss rates induced by the levels of congestion, which are, to some extent, self-caused by their own lack of rate adaptation. It can be argued that, to some extent the users of Internet telephony adapt over longer time scales than those within the *transmission-control protocol*, in the sense that they will defer making calls until the congestion levels are within acceptable bounds.

10.2.4 Pricing

In discussions about the Internet, the question of who pays for it often occurs. Attempts to answer this question typically lead to unconvincing explanations at best and, more usually, to quite a lot of confusion!

Several features can be described. Broadly, there are two types of users of the Internet. The first type is the residential users with access to the Internet through *Internet service providers* over the conventional telephone (or cable-TV) networks. Such users connect through a modem, which necessarily limits their access bandwidth. Access speeds of 56 kbit s^{-1} are now commonly available. The second type of user is more usually found in businesses or academic environments, where a local-area network of computers interconnects with the Internet. Normally, though not necessarily, this interconnection is at a higher bandwidth than that for the residential customers. Although the interconnection is again provided by an *Internet Service Provider*, the contract is with the central management for a group of users, not normally with the individual users themselves.

The standard form of pricing seen by residential customers is a monthly subscription charge (a typical charge being £10 or less) to the *Internet Service Provider* together with any telephony charges associated with the dial-up connection. In the UK, the dial-up connection is normally charged at the local call rate. In the USA, local calls are normally free. Thus, in the UK there is a usage-based component to the user's charge (dependent on the *time* spent connected, though not on the actual *volume* of data transferred), whereas in the USA, there would not normally be any usage-based component to the charge. The phenomenon whereby customers in the USA dial up and stay connected through a local telephone circuit for periods of days at a time is not unknown and has caused much concern to telephone-operating companies in the USA.

Several *Internet Service Providers* in the UK have recently dropped the subscription part of the charge and just rely on receiving a proportion of the local call rate deriving from interconnection payments between the telecom operators.

10.2.5 The example of JANET

The UK academic network (known as JANET) provides a very interesting and topical case study. Since the adoption of IP protocols at the end of the 1980s, the volumes of traffic have grown at a high rate (estimates suggest a trebling each year). Between the various UK academic institutions there has been relatively little difficulty in keeping pace with this growth. The bottlenecks have been the much more expensive transatlantic connections. Here there have been serious difficulties in supplying sufficient bandwidth to meet the demand for resources. Approaches to these problems not only provide much-appreciated relief to UK academic users (of the World Wide Web especially), but have also focused attention on research into the growing multi-disciplinary field of Internet economics, which lies at the heart of this article.

One approach to the problems of handling huge numbers of accesses to Web pages has been the important development of Web-caching systems and services. Another approach has been the introduction of a usage-based charge for traffic in-bound to the UK academic network. This charging began in August 1998 and consists of a *volume* charge of 2 p per megabyte of data transferred (except for a low-charge period of several hours at night, during which it is currently free). The aggregated charge is billed to separate institutions and, for an initial transition period, it is centrally subsidised, reducing the charge from 2 p to 1 p per megabyte. These levels were set in order to recover the shortfall between fixed funding levels and the total cost of providing the necessary bandwidth. (Note that the unit cost of providing the bandwidth is falling, but not at a sufficient rate to compensate within a constant budget for the growth in traffic.) It is currently too early to report other than anecdotal effects arising from these charging mechanisms. Certainly, institutions have begun to monitor more closely their use of network facilities, but, as yet, most standard users seem at best only dimly aware that such charging is taking place. This could easily change very quickly and it will be fascinating to see the variety of responses taken, both by institutions and by users!

10.2.6 Difficulties

There are two major obstacles that are causing fundamental concerns about how well the current Internet will scale up in the future. One concern is the increasing degree of diversity, or heterogeneity, of the applications being used over the Internet. It is no longer the case that the majority of the traffic is confined to just a few well-understood types, such as telnet, file transfer and e-mail traffic. Now, we need to add Web traffic, Internet-telephony traffic, and many more types that few have any doubt will be invented and just as rapidly adopted. The second concern is that, together with this diversity at the traffic level, there is also increasing diversity in value, or utility, terms. Not every packet transferred is equally valued. Some packets in an Internet-telephony connection may well be lost without the users noticing, whereas data being lost in a file transfer would normally trigger re-sending of data. The value attached to the data does not just depend on the application that generated them, but it will also increasingly depend on the context and the human user's perceptions. Consider how a frivolous browsing of a Web page by one person could correspond to a much desired purchasing decision in some e-commerce transaction for another person. These two concerns, of increasing diversity in types of traffic and users' valuations for quality, lie at the heart of much of the research into the way the Internet will evolve.

10.3 Proposals of the Internet Engineering Task Force

The *Internet Engineering Task Force* is the official body that coordinates research and development for the Internet. Two of its activities are directed towards tackling some of the problems raised in the previous section.

The proposals under the umbrella of *integrated services* concern the introduction of various classes of service describing a user's connection. The various classes of service would then, potentially, be handled in different ways appropriate to their characteristics. One class would correspond to the existing best-effort behaviour, whereas another would provide statistical guarantees of the quality of service seen by the connections within that class.

A second approach under investigation by the Internet Engineering Task Force is known as *differentiated services*. In this approach, the packets sent across the Internet are labelled or tagged with some priority information. This might take the form of just one or two extra bits of information added to each packet's header in order to describe whether the network should regard this packet as being of high or low priority (or somewhere in between). High-priority packets might be processed through queues in the routers that take precedence over the queues for low-priority packets. In this way, real-time voice communications would not see the congestion that low-priority e-mail or Web traffic would otherwise cause.

Both of these proposals strive to solve the difficulties of allocating scarce resources among competing heterogeneous users with diverse notions of quality.

10.4 Proposals for pricing

We now turn to some of the proposals for pricing that have recently been made within the emerging field of Internet economics.

10.4.1 A smart market

Seminal work by MacKie-Mason and Varian (1995) described a *smart-market* approach to allocating scarce network resources to the users who value them the most. The approach can briefly be described as follows. Each user adds a value attribute to their packets that specifies the user's willingness to pay for the transportation of this individual packet through the network. At individual resources within the network the various com-

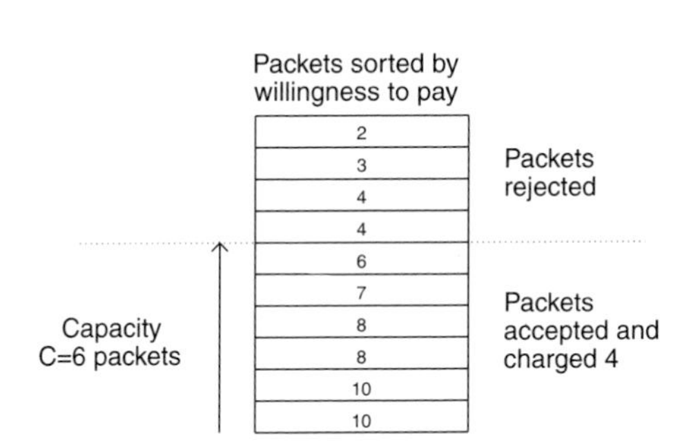

Figure 10.1. A smart market. This figure shows ten packets simultaneously offered to a resource with a capacity for six packets. The willingness to pay ranges from 2 to 10 units. Packets valued 6 and above are accepted while those valued 4 or below are rejected. The accepted packets are all charged an amount of 4 units.

peting packets are sorted according to their willingness-to-pay attribute. The resource then drops all those low-valued packets that lie beyond its capacity threshold. In order to deter users from simply attaching higher and higher willingness-to-pay values to their packets, each user is charged when their packets are carried by the resource. The amount they are charged is not their actual willingness to pay but the willingness to pay of the highest-valued packet dropped. The economic language to describe this procedure is to say that the resource conducts an *auction* for the available bandwidth. The choice of the charge that each user is required to pay when their packet is carried has been shown to be such as to provide an incentive for users to choose their true willingness to pay when they send their packets. Figure 10.1 shows an example with ten packets offered to a network resource whose capacity is only six packets.

This *smart-market* approach, which uses the tools of economic theory for its definition, has been a source of much inspiration to the engineering community wrestling with the interrelated objectives of achieving efficiency of the network together with ease of implementation. Conducting an auction on a packet-by-packet basis within a router (observe that this requires a computationally expensive sorting operation) is not considered viable to implement as such. Its importance lies in describing the goal that more easily implemented approximate schemes must seek to achieve.

10.4.2 Paris Métro pricing

One very pragmatic proposal has been made by Odlyzko (1997). It resembles the pricing used at one time on the Paris Métro. The proposal is to segregate networks into two subnetworks and to charge two different prices. Capacities being equal, the users who are willing to pay more for better quality would choose the high-price network, for which bandwidth should be less scarce (and, hence, connections will be less congested), since the remaining users with lower willingness to pay are deterred from entry. Notice that, just like the first-class carriages on the Paris Métro, there is no technical difference between the subnetworks. The improvement in quality is achieved by the natural behaviour of the self-interested users rather than by any clever or complex actions on the part of network resources to respect priority marks on the packets or service classes.

This proposal has many merits and is likely to be the source of many further investigations to determine its eventual role. One such investigation by Gibbens *et al.* (1998) considered the way in which schemes of this type would be affected by competing network suppliers. Would both the competing networks wish to operate a Paris Métro style of pricing or would they collapse to head-to-head competition on price alone? The initial conclusions from these studies suggest that Paris Métro pricing would not be expected to emerge in a competitive market situation. However, much work remains to be done in this area to determine the extent to which these conclusions remain valid under less-severe modelling assumptions.

10.4.3 Packet marking

The final pricing proposal considered in this article results from the work of Kelly and co-workers (Kelly 1997, Kelly *et al.* 1998, Gibbens and Kelly 1999 a, b). In this proposal, users adapt their sending rates much as they do in the existing *transmission-control-protocol* algorithms of Jacobson. However, there are two differences from the existing framework. The first difference is that the network provides a more refined notion of congestion than simply that of the dropping of packets. The resource feeds back a measure of the true congestion: a quantity that mathematicians and economists call the *shadow price* for the resource. The second difference is that the users no longer need to behave identically when faced with congestion information. Instead, they are not just allowed but indeed *encouraged* to follow their own interests in the way that they adapt the rate of sending packets into the network. Using the principle that the users are charged the shadow price for congestion, the users will have both the knowledge and the incentive to vary their behaviour in such a way as to use the network most efficiently. The overall system architecture of users and resources is shown in Figure 10.2.

Thus the network has an additional job to do and the users have greater

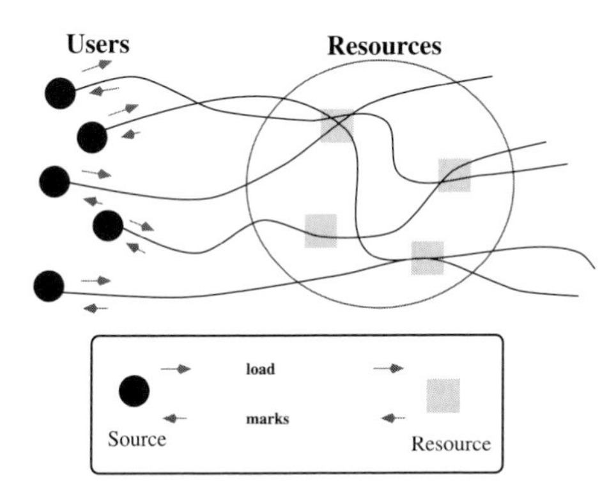

Figure 10.2. Packet marking. This figure shows the overall system architecture of users and resources. The users produce packets that act as load for resources along a route to the destination node. Each resource receives the aggregated load from all users whose routes traverse the given resource and selectively marks packets in order to indicate the level of congestion (shadow price). The marks are returned to the users, who then adjust their rates of sending packets in accordance with their own interests.

freedom in their choice of actions. An important contribution of the research on this topic has been to determine how difficult or complex these jobs are for the designers of network routers and for the software developers of the future seeking to provide integrated multi-media applications at lowest possible (network) cost.

For the network resources, their job can be accomplished by setting marking information on the packet that is conveyed back using the same acknowledgement procedure as that which the *transmission control protocol* uses to indicate receipt of packets. The average rate of packet marking would then signify the shadow price. Research into the appropriate algorithms for choosing which packets to mark and which packets to leave unmarked is still continuing but early work shows that very simple strategies on the part of the routers, requiring minor extra overheads (certainly compared with, say, those for sorting packets), can produce adequate descriptions of the shadow prices.

Users' strategies have also been the subject of much investigation. Research has shown that, combined with use of shadow-price information, it is possible to make minor modifications to the *transmission-control-protocol* algorithms that lead to dramatically different allocations of resources among connections to users. In particular, users can be responsive to the congestion levels in a way that leads to the resources being allocated according to the expressed willingness to pay, whatever those values actually happen to be. Thus, a truly differentiated network of services can be formed without the network's resources having to understand or interpret any classifications made by the users (or by network designers acting on their behalf).

10.5 Future developments

Using the packet-marking framework of the previous section, it is possible to build more sophisticated notions of quality than simply that of the packet-dropping rate. An example is provided by Gibbens and Kelly (1999a), who show how it would be possible to develop a gateway mechanism that effectively translates between the world of packets and marks (which incur a fixed, possibly notional, charge per mark) and the world of non-adaptive sources (such as the current Internet telephony users, who prefer to be told whether their entire connections are accepted or blocked). The gateway could be an extra component that monitors the level of congestion within the network (by means of observing the process of marks flowing back from the resources) and takes decisions concerning the acceptance or rejection of requests from Internet-telephony users. It functions as a form of risk taker on behalf of the users. The gateway is really a virtual device and need not be constructed as a separate component in hardware but can, instead, be implemented as a distributed-software process running at a variety of convenient locations.

The approach outlined in this paper describes a public network together with a system for micro-payments. It is worth remarking that, within a private network of cooperative users, the feedback of shadow-price information to the end systems using packet marking might alone be sufficient to allow the efficient operation of the network's resources.

In a public network of non-cooperative users, the addition of micro-payments would appear to be essential in order to provide the necessary incentives. There has been some speculation that once such a payment system were in place, then *Internet Service Providers* would have a natural means by which to bill for a range of further services, including content, and a whole system for e-commerce would follow.

Concern about the potential transaction costs associated with micro-payments has led to interesting suggestions for probabilistic payment methods for use in the case in which there is a large number of very small transactions.

It is hard to speculate with too much certainty about future developments of the Internet, but it would seem reasonable to suggest that there will be greater opportunities for advancement if the economic and teletraffic problems are resolved via a system of congestion pricing that is built from simple mechanisms. This would ensure the efficiency, robustness and strength with which to support our communication needs for the start of the new millennium and beyond.

10.6 Further Reading

Gibbens, R. J. and Kelly, F. P. 1999a Distributed connection acceptance control for a connectionless network. In *Proc. 16th Int. Teletraffic Congress, Edinburgh, June 1999*, pp. 941–52. http://www.statslab.cam.ac.uk/~frank/dcac.html.

Gibbens, R. J. and Kelly, F. P. 1999b Resource pricing and the evolution of congestion control. *Automatica* **35**, 1965–85. http://www.statslab.cam.ac.uk/~frank/evol.html.

Gibbens, R., Mason, R. and Steinberg, R. 1998 Multiproduct competition between congestible networks. University of Southampton Discussion Paper on Economics and Econometrics, no. 9816

Jacobson, V. 1998 Congestion avoidance and control. In *Proc. ACM SIGCOMM '88*, pp. 314–29. (An updated version is available via ftp://ftp.ee.lbl.gov/papers/congavoid.ps.Z.)

Kelly, F. P. 1997 Charging and rate control for elastic traffic. *Euro. Trans. Telecomm* **8**, 33–7. http://www.statslab.cam.ac.uk/~frank/elastic.html.

Kelly, F. P., Maulloo, A. K. and Tan, D. K. H. 1998 Rate control in communication networks: shadow prices, proportional fairness and stability. *J. Oper. Res. Soc.* **49**, 237–52. http://www.statslab.cam.ac.uk/~frank/rate.html.

MacKie-Mason, J. K. and Varian, H. 1995 Pricing the Internet. In *Public Access to the Internet*. New York: Prentice Hall.

Odlyzko, A. 1997 A modest proposal for preventing Internet congestion. AT&T research, http://www.research.att.com/~amo/doc/recent.html.

Contributor biographies

J. M. T. Thompson

Michael Thompson was born in Cottingham, Yorkshire, on 7 June 1937 and studied at Cambridge, where he graduated with first-class honours in Mechanical Sciences in 1958 and obtained his PhD in 1962 and his ScD in 1977. He was a Fulbright researcher in aeronautics at Stanford University and joined University College London (UCL) in 1964. He has published four books on instabilities, bifurcations, catastrophe theory and chaos and was appointed professor at UCL in 1977. Michael was elected a fellow of the Royal Society in 1985 and was awarded the Ewing Medal of the Institution of Civil Engineers. He was a senior SERC fellow and served on the IMA Council. In 1991 he was appointed director of the Centre for Nonlinear Dynamics at UCL. He is currently editor of the Royal Society's *Philosophical Transactions* (Series A), which is the world's longest-running scientific journal. His scientific interests include nonlinear dynamics and their applications. His recreations include walking, tennis and astronomy with his grandson Ben, shown below.

Adrian Kent

Adrian Kent graduated from the University of Cambridge in 1981 with a first-class degree in mathematics and went on to take a PhD in Cambridge, at the Department of Applied Mathematics and Theoretical Physics (DAMTP). After postdoctoral work at the Enrico Fermi Institute, University of Chicago, and the Institute for Advanced Study, Princeton, NJ, he returned in 1989 to the DAMTP, Cambridge, where he is currently a Royal Society university research fellow. His research interests include the foundations of quantum theory, classical and quantum cryptography, quantum computing and communication, representation theory and two-dimensional conformal field theory. His hobbies include real tennis, theatre, reading, positive-expectation gambling and watching cricket.

Derek Lee

Derek Lee is a condensed-matter theorist. He is interested in quantum phases of matter whose existence depends on quantum correlations, interactions and disorder. These include disordered superfluids, dirty metals and the cuprate superconductors. He was born in Hong Kong and studied Natural Sciences at Cambridge, where he graduated with first-class honours in 1987 and obtained his PhD in 1990. His postdoctoral work on one-dimensional metals, electron localisation and the gauge theory of unconventional metallic behaviour in high-temperature superconductors was carried out in Oxford and, as a Lindemann fellow and a NATO/EPSRC fellow, at the Massachusetts Institute of Technology, USA (1993–96). In 1997, he joined the Condensed Matter Theory Group at Imperial College, London, as a Royal Society university research fellow.

Andrew Schofield

Andrew Schofield is a Royal Society university research fellow and proleptic lecturer at the University of Birmingham. His research interests include the theory of correlated quantum systems and their manifestation in oxide materials such as high-temperature cuprate superconductors and also in heavy-fermion metals and semiconductor structures. He originates from Surrey, and graduated from Cambridge with a first-class BA and Mott prize in 1989. In his PhD, awarded at Cambridge in 1993, he used a gauge theory approach to study doped Mott insulators. He then worked at Rutgers University, USA, studying Kondo models and developing a phenomenological approach to transport in the cuprate metals. He returned to Cambridge as a fellow of Gonville and Caius College in 1996 and moved to Birmingham in 1999.

William Power

William Power was born in Barnet, Hertfordshire, in 1969. He studied physics at Imperial College, graduating with first-class honours in 1990. He studied for a Masters degree at the University of Rochester, USA, between 1990 and 1992 and then did research into laser–atom interactions to obtain his PhD from Imperial College in 1995. After that, he was awarded a European science exchange fellowship from The Royal Society, which allowed him to work for a year on atom and quantum optics at the Universität Konstanz in Germany. William has recently been investigating the possibility of using atom interferometry to look for quantum effects in gravity at Queen Mary and Westfield College in London and has been working as an analytical consultant for Analyticon Limited. He now lives in New Zealand.

Giles Davies

Giles Davies studied at Bristol University, graduating with first-class honours in chemical physics in 1987. He obtained his PhD in 1991 in the Cavendish Laboratory Semiconductor Physics Group of Cambridge University, having been elected to a Senior Rouse Ball Studentship at Trinity College. Giles was awarded an Australian Research Council post-doctoral fellowship and joined the Australian National Pulsed Magnet Laboratory in Sydney, where he worked for three years. In 1995, he returned to the Cavendish as a Royal Society university research fellow. He was elected Trevelyan Fellow at Selwyn College and, aged 33, is now College Lecturer and Director of Studies in Physics and sits on the College Council. His scientific interests include the optical and electronic properties of low-dimensional electronic and biomolecular systems.

Michael Ziese

Michael Ziese was born in 1964 in Uetersen, Schleswig-Holstein, Germany, and studied at the Universität Hamburg, where he graduated with a diploma in physics in 1992. He obtained his PhD *summa cum laude* from the Universität Bayreuth in 1995 with a study on flux-line lattice dynamics in superconductors. After post-doctoral work at the Vrije Universiteit Amsterdam in 1995 and the Universität Leipzig in 1996, he joined the magnetism group at Sheffield University as a research associate in January 1997. In October 1999 he joined the Physics Department at the Universität Leipzig in order to undertake research for his Habilitation. His scientific interests include transport in oxides, especially colossal magnetoresistance materials and high-temperature superconductors. His recreational activities include reading books and going on family walks in the countryside.

Ifor Samuel

Ifor Samuel was born in London and read natural sciences at Cambridge, graduating in 1988. He remained there for his PhD on femtosecond spectroscopy of conjugated polymers and was elected a fellow of Christ's College in 1991. He spent two years at CNET–France Telecom in Paris working on nonlinear optics of organic material. This was followed by five years in Durham as head of the light-emitting-polymer research group before his current position of Professor of Physics at the University of St Andrews. His current research involves the use of optical and electrical spectroscopy to study semiconducting polymers with the aim of understanding and improving them. He is now aged 33 and a Royal Society university research fellow. His other interests include music, theatre, travelling and hill-walking.

Michele Mosca

Michele Mosca was the Robin Gandy Junior Research Fellow at Wolfson College, Oxford from 1998 to 2000. He obtained a Bachelor of Mathematics from the University of Waterloo, Canada, in 1995 (alumni gold medal winner), and an MSc in Mathematics and the foundations of computer science (with a distinction) from the University of Oxford, where he also completed his DPhil on quantum computer algorithms. He has made major contributions to the unification of quantum algorithms and is a co-inventor of the polynomial method for studying the limitations of quantum computing. Together with collaborators at Oxford, he implemented some of the first quantum algorithms. He is continuing work on quantum computation as Assistant Professor of Mathematics at the University of Waterloo. His other interests include rowing and languages.

Richard Jozsa

Richard Jozsa was born in Melbourne, Australia, and studied at Monash University, graduating with first-class honours in Mathematics. He obtained his DPhil in 1981 from Oxford University. Throughout the 1980s he held postdoctoral positions at Oxford and at McGill, Sydney and other universities in Australia. Richard began working in quantum computation in 1989. In 1992, with David Deutsch, he gave the first demonstration of the power of quantum computing over classical computing. In 1994 he co-invented quantum teleportation. In 1996 he was a Royal Society Leverhulme senior research fellow and subsequently became Professor of Mathematical Physics at the University of Plymouth. Currently Richard is an EPSRC senior research fellow and Professor of Computer Science at the University of Bristol. Recreations include gastronomy and playing the violin.

Andrew Steane

Andrew Steane obtained a doctorate in physics from Oxford University in 1991, in the area of experimental atomic physics. He was a junior research fellow at Merton College, Oxford, and then a European research fellow at the Ecole Normale Supérieure, Paris, where he developed experiments in laser cooling and interferometry of atoms. From 1995 to 1999 he held a Royal Society university research fellowship at the Clarendon Laboratory, Oxford, and since then has been a Lecturer and Tutorial Fellow of Exeter

College, Oxford. Through studying the theory of quantum interference involving many particles, he established the basic principles of quantum error-correction theory. He was awarded the 2000 Maxwell Medal and Prize for this work. He now combines this research with experiments on trapped ions. He is now aged 35, married and lives in north Oxford. He leads a childrens' group at his church and is a keen walker and singer.

Artur Ekert

Artur Ekert is a Professor of Physics at Oxford University and a Fellow and Tutor of Keble College, Oxford. He obtained his DPhil from Oxford in 1991. In his doctoral thesis he introduced entanglement-based quantum cryptography. From 1991 until 1998 he was a Research Fellow at Merton College, Oxford, and in 1993 he was elected the Royal Society Howe Research Fellow. Since 1992 he has been in charge of the Quantum Computation and Cryptography Research Group (which was recently turned into a research centre) in the Clarendon Laboratory, Oxford. He has been a consultant on quantum information technology (both to industry and to government agencies). For his work in quantum cryptography he was awarded the 1995 Maxwell Medal and Prize. He enjoys outdoor sports.

Standing (from left to right): Artur Ekert, Andrew Steane, Michele Mosca. Seated: Richard Jozsa.

Russell Cowburn

Russell Cowburn was born in Newcastle-upon-Tyne and, after a year working in the defence electronics industry, came to Cambridge in 1990 to study natural sciences. He graduated in 1993 with first-class honours and immediately began a PhD at the Cavendish Physics Laboratory, Cambridge. His research into the magnetism of films of atomic thickness led him to spend a year working at the CNRS, Paris. In 1997 he was elected to a research fellowship at St John's College, Cambridge to carry out research into magnetism and nanotechnology. He moved to Durham University in 2000 to take up a University Lectureship. He is aged 29 and has published over 25 papers. He is married and a committed Christian. His interests include hill walking, Brahms and P. G. Wodehouse.

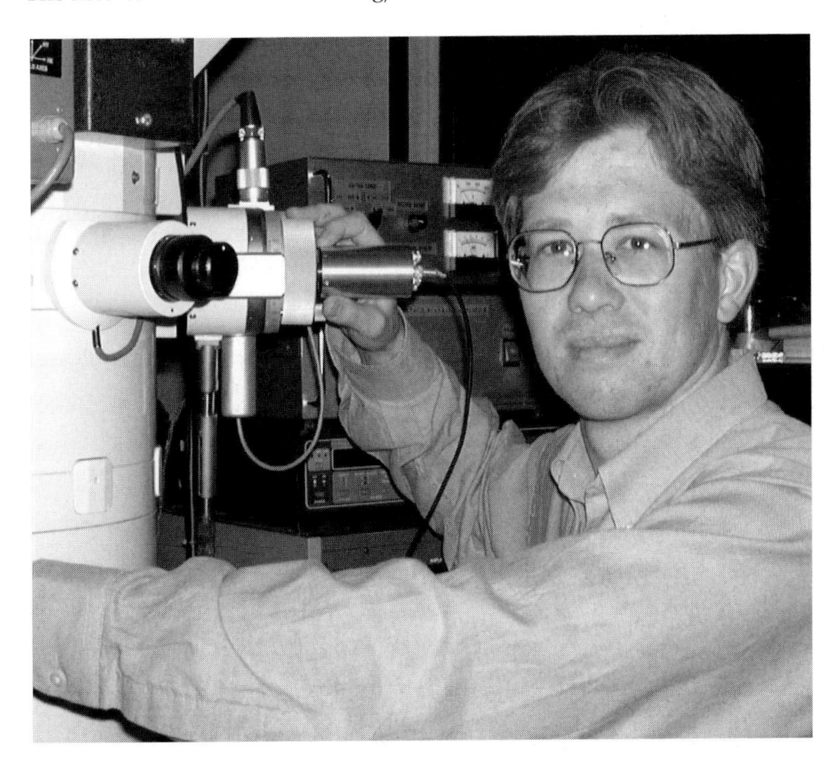

Polina Bayvel

Polina Bayvel, now aged 34, was born in the USSR. She received her BSc(Eng) and PhD degrees in Electronic & Electrical Engineering from University College London (UCL) in 1986 and 1990, respectively. In her PhD work she specialised in non-linear fibre optics and its applications. Following a Royal Society post-doctoral exchange fellowship in the Fibre Optics Laboratory at the General Physics Institute in Moscow (USSR Academy of Sciences) in 1990, she worked as a Principal Systems Engineer at STC Submarine Systems Ltd (Greenwich, UK) and BNR (Bell–Northern Research – subsequently Nortel Networks) Harlow UK and Ottawa, Canada, on the design and planning of high-speed optical transmission networks.

In December 1993 she was awarded a Royal Society university research fellowship at the Department of Electronic & Electrical Engineering, UCL. This has been extended until 2003, concurrently with an appointment to a lectureship in 1966, senior lectureship in 1998 and a readership in 2000 in the same department. Since 1994 she has built up and heads the Optical Networks Group whose research activities are in the areas of high-speed optical communication systems, networks and associated devices. She has authored/co-authored over 80 refereed journal and conference papers. She enjoys languages, reading, theatre, winter sports and hiking.

Richard Gibbens

Richard Gibbens was born in 1962 in London. He studied mathematics at the University of Cambridge, graduating with an MA in mathematics, a diploma in mathematical statistics and a PhD in 1988 in the Statistical Laboratory. His doctoral thesis work, a collaboration between the University of Cambridge and British Telecom, was on the design and analysis of dynamic routing strategies in telecommunications networks, resulting in the development of the dynamic alternative routing strategy, which is now in operation in the British Telecom trunk network. He has worked in the area of modelling of communications networks, mainly at the Statistical Laboratory, University of Cambridge, but was also a visiting consultant at AT&T Bell laboratories, Murray Hill, NJ, during 1989. He was appointed to a Royal Society university research fellowship in 1993.

Index